AF818850

Contributions to Phenomenology

In Cooperation with The Center for Advanced Research in Phenomenology

Volume 105

Series Editors
Nicolas de Warren, Department of Philosophy, Pennsylvania State University, State College, PA, USA
Ted Toadvine, Department of Philosophy, Pennsylvania State University, State College, PA, USA

Editorial Board
Lilian Alweiss, Trinity College Dublin, Dublin, Ireland
Elizabeth Behnke, Ferndale, WA, USA
Rudolf Bernet, Husserl Archive, KU Leuven, Belgium
David Carr, Emory University, Atlanta, GA, USA
Chan-Fai Cheung, Chinese University of Hong Kong, Sha Tin, Hong Kong
James Dodd, New School University, New York, USA
Lester Embree, Florida Atlantic University, Florida, USA
Alfredo Ferrarin, Università di Pisa, Pisa, Italy
Burt Hopkins, University of Lille, Lille, France
José Huertas-Jourda, Wilfrid Laurier University, Waterloo, Canada
Kwok-Ying Lau, Chinese University of Hong Kong, Sha Tin, Hong Kong
Nam-In Lee, Seoul National University, Seoul, Korea (Republic of)
Dieter Lohmar, University of Cologne, Cologne, Germany
William R. McKenna, Miami University, Ohio, USA
Algis Mickunas, Ohio University, Ohio, USA
J.N. Mohanty, Temple University, Philadelphia, USA
Dermot Moran, University College Dublin, Dublin, Ireland
Junichi Murata, University of Tokyo, Tokyo, Japan
Thomas Nenon, The University of Memphis, Memphis, USA
Thomas M. Seebohm, Johannes Gutenberg-Universität, Mainz, Germany
Gail Soffer, Roma Tre University, Rome, Italy
Anthony Steinbock, Southern Illinois University, Carbondale, USA
Shigeru Taguchi, Hokkaido University, Sapporo, Japan
Dan Zahavi, University of Copenhagen, Copenhagen, Denmark
Richard M. Zaner, Vanderbilt University, Nashville, USA

Scope

The purpose of the series is to serve as a vehicle for the pursuit of phenomenological research across a broad spectrum, including cross-over developments with other fields of inquiry such as the social sciences and cognitive science. Since its establishment in 1987, *Contributions to Phenomenology* has published more than 100 titles on diverse themes of phenomenological philosophy. In addition to welcoming monographs and collections of papers in established areas of scholarship, the series encourages original work in phenomenology. The breadth and depth of the Series reflects the rich and varied significance of phenomenological thinking for seminal questions of human inquiry as well as the increasingly international reach of phenomenological research.

All books to be published in this Series will be fully peer-reviewed before final acceptance.

The series is published in cooperation with The Center for Advanced Research in Phenomenology.

More information about this series at http://www.springer.com/series/5811

Thomas Nenon
Editor

Thomas Seebohm on the Foundations of the Sciences

An Analysis and Critical Appraisal

Springer

Editor
Thomas Nenon
Department of Philosophy
University of Memphis
Memphis, TN, USA

ISSN 0923-9545 ISSN 2215-1915 (electronic)
Contributions to Phenomenology
ISBN 978-3-030-23660-1 ISBN 978-3-030-23661-8 (eBook)
https://doi.org/10.1007/978-3-030-23661-8

© Springer Nature Switzerland AG 2020
This work is subject to copyright. All rights are reserved by the Publisher, whether the whole or part of the material is concerned, specifically the rights of translation, reprinting, reuse of illustrations, recitation, broadcasting, reproduction on microfilms or in any other physical way, and transmission or information storage and retrieval, electronic adaptation, computer software, or by similar or dissimilar methodology now known or hereafter developed.
The use of general descriptive names, registered names, trademarks, service marks, etc. in this publication does not imply, even in the absence of a specific statement, that such names are exempt from the relevant protective laws and regulations and therefore free for general use.
The publisher, the authors, and the editors are safe to assume that the advice and information in this book are believed to be true and accurate at the date of publication. Neither the publisher nor the authors or the editors give a warranty, expressed or implied, with respect to the material contained herein or for any errors or omissions that may have been made. The publisher remains neutral with regard to jurisdictional claims in published maps and institutional affiliations.

This Springer imprint is published by the registered company Springer Nature Switzerland AG.
The registered company address is: Gewerbestrasse 11, 6330 Cham, Switzerland

Dedicated to the memory of Lester Embree (1928–2017), tireless disciple and ambassador of phenomenology as a living tradition

Preface

This volume includes papers originally presented at a symposium on Thomas Seebohm's *History as a Science and the System of the Sciences* that was sponsored by the Center for Advanced Research in Phenomenology, organized by Lester Embree and held in Memphis, TN, on March 18–19, 2016. It was the last of many conferences he organized over the years. It also includes several additional contributions by authors who were not able to attend the conference. The task of editing and publishing it fell to the local organizer Thomas Nenon who is grateful to Lester for his idea for the conference and his efforts in making the conference and this volume possible.

Memphis, TN, USA — Thomas Nenon

Introduction

The papers contained in this volume represent the first and up until now only comprehensive effort at a critical analysis and explication of the work of Thomas Seebohm, one of Germany's (and America's) leading phenomenologists and hermeneuticists of the second half of the twentieth century. It focuses especially upon his final work, *History as a Science and the System of the Sciences*, published in 2015, as the culmination and summary of his work in historical and phenomenological investigations into the foundations, nature, and limits of modern sciences, including not just history but the *Geisteswissenschaften* more generally, along with the social and cultural and natural sciences as well. The essays in this volume reflect that range. They include essays describing his basic approach and insights in that work and in *Hermeneutics: Method and Methodology*[1] from 2004. Several of them draw explicit comparisons to other important figures whose work covers similar ground. These include Dilthey, Gadamer, Husserl, and Schutz but also some others whose work is less well-known to most philosophers, such as mathematician Leopold Kronecker and social scientist Felix Kaufmann. All these essays help illustrate Seebohm's unique contributions and situate him in larger discussions within twentieth-century hermeneutics and philosophy of science. Other essays are also dedicated to an analysis of Seebohm's positions on questions about the foundations and methodologies of specific sciences or areas of science such as history, the humanities, the social sciences, or the natural sciences (particularly physics). The net result is an in-depth study and a helpful overview of Seebohm's general approach and his specific views on various areas of modern science. One reason why this is important is that Seebohm's project of providing a systematic phenomenological analysis of the philosophy of modern science and the sciences as a whole is an important desideratum within the phenomenological tradition and one that he has pursued in more depth and breadth than perhaps any other thinker within that tradition. As such, his work is of interest not just as a matter of historical scholarship but also and above all as an important contribution to phenomenology and to the

[1] Other earlier book-length studies in this area also include Seebohm (1972) and (1977).

philosophy of science and the sciences as such, one that deserves attention by scholars from any philosophical tradition and from scholars within these areas who are interested in thinking about the foundations of their disciplines.

A few brief remarks to Seebohm himself and his work are perhaps in order before turning to the essays in this volume. Thomas Seebohm, born in 1934 in Gleiwitz in Upper Silesia (now part of Poland), was one of Germany's premier phenomenologists. His interests and expertise extended well beyond phenomenology in the narrow sense, to hermeneutics, formal logic, and analytic philosophy as well. Of course, as this volume amply documents, he was interested in the theory and philosophy of a whole range of scientific disciplines from the natural and social sciences to formal sciences and the human sciences, with a special interest in the philosophy of history. In addition to his philosophical interests, he was also a recognized authority on Slavic literature and history as well, especially medieval Russian philosophy and culture.

Before beginning his university studies, he had learned a trade (cabin-making). Having studied philosophy, Slavic languages and literatures, and sociology in Hamburg, Bonn, and Saarbrücken, he completed studies in 1960 in Mainz under the famous Kant scholar and hermeneuticist Gerhard Funke, with a dissertation *Die Bedingungen der Möglichkeit der Transzendentalphilosophie*: *Edmund Husserls transcendental-phänomenologischer Ansatz, dargestellt im Anschluss an seine Kant-Kritik* (*The conditions of the possibility of transcendental philosophy: Edmund Husserl's transcendental-phenomenological assessment, presented in connection with his criticism of Kant*), which was later published with Bouvier Verlag in 1962. His first regular appointment was as an assistant in Mainz beginning in 1965, followed by a visiting professorship at Penn State in 1970 that would eventually become a permanent position that he occupied from 1973 to 1984, before returning finally to Mainz where he became Funke's successor to the Chair there, where he remained until his retirement in 1999. The extensive list of his publications can be found at the end of this volume, but it should be noted that, in addition to his publications, his teaching and engagement in philosophical organizations devoted to phenomenology in German and North America made him one of the most influential figures in phenomenology and hermeneutics of his generation.

The first group of papers that concentrate on specific sciences or areas, especially the human and social sciences, begins in Chap. 1 with David Carr's discussion of Seebohm's position on the system of sciences in general and history's place within it before turning to what Carr calls the "paradox of subjectivity" that arises at the latest with Husserl's anchoring of the sciences in a subjective "attitude"[*Einstellung*] that is the basis for scientific objectivity, which Carr sees as best addressed by recognizing that the tension is best framed not as a difference between the personalistic versus the naturalistic attitude, but as the difference between the everyday, non-reflective, "mundane" attitude and that of philosophical or specifically phenomenological reflection. In Chap. 2, Thomas Nenon analyzes Seebohm's thoughts on the various kinds of *interpretation* that form the foundation for what in the English-speaking world have come to be called "the humanities." Here, one important note is the difference between the way that different projects and even disciplines in the

United States are clustered together in academic departments in a way that is quite different from Seebohm's description of philology and history, which still fits much more closely with the self-understanding of the units within the *Geisteswissenschaften* that we think of as corresponding to the humanities in the United States and most English-speaking countries. It is not surprising that Michael Barber's analysis of Seebohm on the cultural sciences in Chap. 3 could have just as easily been included among the comparative studies, since it makes perfect sense to use a comparison with Alfred Schutz's work in this area as a starting point, and it represents the most extensive and careful application of Husserlian phenomenological insights and approaches to the social sciences and their foundation and methodology. All the more significant, then, that Barber, who is unusually familiar with and sympathetic to Schutz's project, is able to identify important respects in which Seebohm is able to add to what Schutz has contributed to this field.

A second group of essays introduce and evaluate Seebohm's work through comparative analyses. In Chap. 4, Robert Dostal continues a personal and scholarly dialogue with Seebohm over several decades about Seebohm's position in relation to the other most important hermeneuticist in post-war Germany, Hans-Georg Gadamer. He argues that their differences are less contradictions than differences in their basic projects and interests, with Gadamer's leading concerns being ontological and Seebohm's epistemological, a point Dostal argues that Gadamer would accept. The result is not just illuminating for Dostal's analysis of Seebohm but also casts an original and instructive light on Gadamer's project and the hermeneutic project as well. Chapter 5 by Rudolf Makkreel continues the discussion of Seebohm's hermeneutics through a comparison with Dilthey. Makkreel provides a very careful and nuanced comparison of the differences in how they describe various kinds and levels of understanding, showing the precise points of departure that result in different conceptions and different notions of the nature and foundations of the social sciences. His critical conclusion ends up affirming a more pivotal notion of understanding than Seebohm's in claiming that understanding is not merely the primary method of the human and social sciences, as Seebohm sees it, but also its envisaged end product, placing it on a higher level than the knowledge that Seebohm sees as the ultimate aim of all sciences, including the social and human sciences. In Chap. 6, Roberto Walton shows how Seebohm builds upon and expands the notion of reduction famously introduced by Edmund Husserl through his description of the "abstractive reductions" at work in the transition from the lifeworld to sciences in general and then the further reductions that are necessary for the constitution of modern natural science. Here, Walton recognizes important continuities between Seebohm's enterprise and the general outlines of such a project that Husserl points to in his *Crisis* volume (1970). Lester Embree's essay in Chap. 7 continues this theme, comparing the abstractive reductions more closely to those at work in Schutz and in Dorion Cairns. In Chap. 8, James Dodd uses a comparison between Seebohm and Kronecker's work in the philosophy of mathematics as a way of framing the broader significance of Seebohm's insights into the emergence of the formal sciences from the lifeworld.

The next group of papers takes up themes from Seebohm's work and applies them to a wider range of problem in the philosophy of science. In Chap. 9, Pedro Alves discusses Seebohm's thoughts on the foundations of the natural sciences and uses this as a fruitful starting point for a critical evaluation of the "Copenhagen School" within modern physics, which claims that the developments of modern quantum physics at the latest have revealed the shortcomings of traditional theories of science from Kant up through phenomenology. He makes a forceful case for the fact that these developments in modern physics have rendered the traditional questions about the foundations of modern physical science anachronistic or moot only under certain interpretations of the developments within twentieth-century physics and that, to the contrary, the kinds of reflections and investigations undertaken by Seebohm, as one successor to Husserlian phenomenology, remain just as relevant as ever.

Marina Banchetti's topic in Chap. 10 is also a comparison between Husserl and Seebohm with a critical analysis of the project of a phenomenological analysis of part/whole relationships, mereology as a foundational project in phenomenology, which she ends up arguing needs some radical reconsideration in light of developments in science subsequent to Husserl's original formulation of the project. The last chapter in this group from Harry Reeder uses a comparison between Seebohm and the social scientist Felix Kaufmann as a way of introducing Kaufmann, a fascinating figure in his own right who was both a member of the Vienna Circle and a phenomenologist to a wider philosophical audience, and to locate Seebohm's positions within the broader framework of discussions in the philosophy of the sciences that were so central to both the analytical tradition and—at least in its inception—the phenomenological tradition, too.

Chapters 12 through 15 present four essays by Seebohm that illustrate the breadth of his philosophical concerns and expertise. Chapter 12 is an analysis of Husserl's remarks on the human sciences in the *Ideas II* and *Ideas III* that expands on the brief critical comments Seebohm makes in the *System of the Sciences* book on Husserl's treatment of that topic. It shows that he studied what remarks Husserl did make carefully and with a critical eye to what he misses as well as the general direction of that work, which Seebohm sees as pointing in the right direction. Chapter 13 is a good example of an original and thoughtful phenomenological interpretation of three key historical figures on the question of the grounding of the basic structures of knowledge in subjective cognition. Seebohm argues that Husserl and Fichte both work toward something like a deduction of those basic structures that is very different from what the literature on Kant took him to be doing, but that Seebohm argues is much closer to what Kant was actually doing in the two deductions from the *Critique of Pure Reason*. This phenomenological approach allows Seebohm to recognize affinities that are more basic than the differences in these thinker's respective terminologies and initial starting points. Chapter 14 continues the discussion of the ultimate grounding of philosophy by tracing out what Seebohm calls the "paradox of subjectivity" in Husserl, one that leads to the criticisms leveled by the early Heidegger, and Husserl's ultimate solution to the paradox, which avoids the charges Heidegger levels against him. It also shows how the problem of an "absolute grounding" for philosophy still remains for both Husserl and Heidegger. Chapter 15 then rounds out this group of essays by showing how the same approach can yield surprising results when applied to contemporary figures as well, in this case, Jaakko Hintikka, who comes almost exclusively from the analytic tradition, and J. N. Mohanty, an expert in phenomenology and comparative philosophy who is familiar with the analytical tradition as well. At bottom, Seebohm shows that the real issue is the nature of intentionality. The outcome of applying viewpoints taken from a phenomenological semantics on the level of predication to problems connected with the "possible worlds semantics" as considered in the analytical tradition was that the "world" as it is to be understood in the context of formalized logic cannot be understood as some kind of concrete world, the real world in which we live, the "worlds" in quasi-positionality or worlds constructed in metaphysical speculations as correlates of other intellects. Phenomenological reflection shows how they are the result of a peculiar abstraction and idealization which separates them in principle from all other conceptions of world. Thus, the paradoxes which are connected with the attempts to connect the different concepts of "world" vanish when reflections on the genetic origins of these kinds of abstractions are undertaken.

The volume closes appropriately with a bibliography of Seebohm's extensive publications over a very productive career. The hope is that the essays in this volume will stimulate the reader to explore some of those monographs and essays more closely and with great philosophical gain.

One very important note in closing: This volume and, indeed, the publication of Seebohm's culminating achievement in the *History as a Science and the System of the Sciences* are due to the efforts of Lester Embree, whose contributions to preserving the legacy of the New School and the professors he met there have been unflagging from the time he entered the New School until his death in January 2017. The fact that the philosophical world has become aware of the phenomenological contributions of Aron Gurwitsch, Alfred Schutz, and Dorion Cairns is due to the extensive work in editing and publishing their works and encouraging and supporting papers and conferences on their work. In this case, too, Embree remained devoted to the New School professor and his friend Thomas Seebohm in managing the editing and publication of his final work and in organizing a conference on this work in March 2017 as the last of many phenomenological conferences he organized in over four decades of organizational activities.[2] This volume is dedicated to his memory.

References

Husserl, Edmund. 1970. *The crisis of European sciences and transcendental phenomenology*. Trans. D. Carr. Evanston: Northwestern University Press.

[2] For a more complete account of the many ways in which Embree contributed to fostering phenomenology as a continuing and important intellectual movement into the twenty-first century, see Nenon (2010).

Nenon, Thomas. 2010. Advancing phenomenology as a practical enterprise, In *Advancing Phenomenology: Festschrift for Lester Embree*, 467–464. Dordrecht: Springer.
Seebohm, Thomas M. 1972. *Zur Kritik der hermeneutischen Vernunft*. Bonn: Bouvier.
Seebohm, Thomas M. 1977. *Ratio und Charisma. Ansätze zur Ausbildung eines philosophischen und wissenschaftlichen Weltverständnis im Moskauer Rußland*. Bonn: Bouvier.
Seebohm, Thomas M. 2004. *Hermeneutics: Method and methodology*. Boston: Kluwer Academic.
Seebohm, Thomas M. 2015. *History as a science and the system of the sciences: Phenomenological investigations*. New York: Springer.

Contents

Chapter 1
On Thomas Seebohm's *History as a Science and the System of the Sciences*

David Carr

Abstract In this paper I do three things: I comment on the idea of the "system" of the sciences in relation to Husserl's work and the role of history within that system; I talk about Seebohm's approach to history as a science; and I evaluate his idea of the relation between history and the other sciences with respect to his take on the "paradox of subjectivity."

Here's what I want to do in this paper:

(1) Comment on the idea of the "system" of the sciences in relation to Husserl's work and the role of history within that system
(2) Talk about Seebohm's approach to history as a science
(3) Evaluate his idea of the relation between history and the other sciences with respect to his take on the "paradox of subjectivity."

(1) What Seebohm is doing here is to develop an idea that seems an obvious outcome of Husserl's philosophy but was not carried out by Husserl himself, at least not in any detail. As Seebohm says in the preface, Husserl "offers a general theory of knowledge and philosophy of science, [but] says almost nothing about a phenomenological epistemology of the sciences and a system of the sciences, i.e., what is missing are critical analyses of the *methodologies* of the sciences" (Seebohm 2015, vii). Thus, what Seebohm has done is not an interpretation or a reconstruction of Husserl but an extension of Husserl.

But it also surprises, as the title already tells us. "History as a Science" is the centerpiece. The closest Husserl comes to a system of the sciences (as far as I know) is what we find in *Ideas II*, called studies in the phenomenology of constitution. Part one is devoted to the constitution of material nature, part two to the constitution of animal nature, and part three to the constitution of the "*geistige Welt*," the spiritual world. The three sections thus correspond to the natural sciences, the psychological sciences (roughly), and the humanities. This order, of course, is exactly what we

D. Carr (✉)
New School for Social Research, Emory University, Atlanta, GA, USA
e-mail: dcarr@emory.edu

© Springer Nature Switzerland AG 2020

T. Nenon (ed.), *Thomas Seebohm on the Foundations of the Sciences*, Contributions to Phenomenology 105,
https://doi.org/10.1007/978-3-030-23661-8_1

might expect from the philosophers of Husserl's time. Nature is the underlying basis, and the psychic and human worlds are built upon it. Dilthey took a similar approach, and it bears traces of Hegel's encyclopedia before that. These philosophers opposed the positivist attempt to reduce the human or the psychic to the natural, and so proposed that the human world had its own structure which called for separate sciences with their own conceptual resources. But the natural world was conceptually self-enclosed and could be left alone. The natural sciences were, after all, the great success story of the modern period, and they earned the admiration that philosophy lavished upon them. In relation to the natural sciences, philosophy is reduced to the handmaidenly role of asking after the conditions of their possibility, of examining their basic concepts, which were taken to be universal, necessary, and immutable.

The story was altogether different for the human sciences. The very idea of history as a science was fairly new. Only since the 1810s had history been established in the universities as an academic discipline with its own standards of rigor and method. Toward the end of the century, Dilthey proposed a "critique of historical reason," seeking a parallel to Kant's analysis of the natural sciences. He and the neo-Kantians and others who turned to this problem found it hard to agree on what was essential to the historical sciences and what makes them sciences. The distinctions between explanation and understanding, and between ideographic and nomothetic sciences, were introduced in the effort to clarify how human differed from natural sciences. But this left two problems intact: one was the place of the human world within nature, given the inviolate status of natural science, and the other was the nature of human sciences considered as science or a system of sciences. Are there multiple human sciences, such as psychology, sociology, law, archeology? Is history the "master science" governing them all? How do they all fit together?

These are questions that concern history as a science, but they do not even begin to touch upon the relations between the historical realm and the natural. It turns out that in the course of Husserl's career his conception of those relations underwent an important change, especially when Husserl recognized the historical character of our conception of nature. As for Seebohm, what he proposes here is to start with the historical human sciences, then turn to the natural sciences, and finally to what he calls the systematic human sciences, i.e., psychology and the social sciences, at the end.

(2) But more of that later. For the moment I want to focus on Seebohm's treatment of the historical sciences and their methodology. For him "the task of history and its methods is to reconstruct the past reality of a more or less foreign cultural lifeworld. A historical reconstruction of past reality is, therefore, per se an interpretation of a past reality. The presently given empirical materials for such reconstructions are texts, monuments, and artifacts. Such materials are the *facts for historical research.* Historical reconstructions presuppose, hence, the methodologically guided interpretations of texts, monuments, and artifacts, i.e., history as a science presupposes methodological principles of philological and archaeological interpretation" (124).

Thus, any methodological consideration of history leads us back to philology and hermeneutics as the science of interpretation. Much of Seebohm's first chapter on the methodology of the historical human sciences is thus devoted to history as a science of interpretation, and to the history of hermeneutics. With its background in biblical and classical scholarship, this history emerges in the early nineteenth century in Schleiermacher and Boeckh and is then taken up by Droysen and Dilthey and passed thence to Heidegger and Gadamer. The chief canon of interpretation that emerges in this history is what has come to be known as the hermeneutical circle, the principle of the interdependence of whole and parts. The question of whether the author's intention is key to understanding a text emerges as a major point of contention along the way. (This question is the basis for Seebohm's longstanding *Auseinandersetzung* with Gadamer.)

The connection between the science of interpretation and that of historical reconstruction is a bit more complicated than it appears. It is not as if the philologist first interprets the texts or artifacts and then delivers them to the historian to complete the next step. For one thing the task of interpretation itself presupposes history, since understanding the text as a whole means placing it in its historical context. In this sense philology presupposes history, even as history presupposes philology. Thus, the hermeneutical circle replicates itself at a different level. Furthermore, the science of interpretation has its own history, as Seebohm shows in this section, and is thus part of the history whose job it is to understand.

In any case, history is not finished when it reconstructs "a past reality" or a past lifeworld. Using the materials at its disposal in the present ("facts for historical research") it seeks to arrive at historical facts, and historical facts are reported facts. But reports must not be accepted naively, but rather must be subjected to critical analysis. This means among other things comparing them with other reports and other evidence in order to arrive at the best possible account of what happened. Historical facts are woven into historical narratives that incorporate facts and their developments. But such narratives must also establish causal connections. This leads Seebohm to a second chapter on the methodology of historical human sciences dealing with causal explanations in history.

The bulk of this chapter is devoted to the obvious inadequacies of various attempts in analytic philosophy to treat causal explanation in history and to reduce it to a model that resembles causal explanation in the natural sciences. It is noteworthy that while Seebohm comments on some of the early versions of applying causal explanation to history (Nagel, Popper) he does not go into some of the reactions to this approach by, among others, William Dray and Collingwood, who urge a kind of "explanation by reasons" (basically rational reconstruction) that is a kind of return to the "understanding/explanation" debate. In any case, at the conclusion of this chapter Seebohm reminds us that "history as a science has to locate past realities within a common formal temporal and spatial framework in which they are all given together in a universal context" (153). This means that the astronomy and geography of the time-period in question must be understood as the background horizon. At the same time, it is inevitable that present conceptions of space, time, and geography serve as the background horizon of the present-day historian. This is especially true when historical research takes us in the direction of the pre-history of literate societies in pre-literate cultures. "Research in this direction cannot be restricted to archaeological interpretation of artifacts and monuments. It will have to use increasing amounts of traces that belong to the material of paleontology. The history of humankind, and with it the ontological region of the human or cultural sciences, appears in this dimension as an extension of the history of nature" (156). "It is, hence, impossible to discuss epistemological questions of the human sciences in splendid isolation without considering the fields in which the human and the natural sciences overlap." Thus, "a strict separation of the sciences of 'understanding' and the sciences of 'explanation' does not hold water" (156). The leads, then, to the third section of the book, devoted to the methodology of the natural sciences.

I won't comment on these chapters (7 and 8) on natural science, but it should be noted that Seebohm returns to and expands on this theme of the relation between historical and natural science after he completes these chapters, in chapter 9. He points out that the standard neo-Kantian division between explanation and understanding leads to a number of problems. First, it leaves no place for a history of nature, that is, a historical science of nature, even though natural history, and even astronomical history, are perfectly respectable and long-standing disciplines. Second, it really has no place for the role that genuine causal explanations have within human-historical research. Third, it does not allow for the "interplay of methods" that this situation brings along with it. The key is that such disciplines as natural history are ideographic, in that they track particular developments rather than search for laws, but they are not interpretative or hermeneutical, since they do not deal with "fixed life expressions." Examples are geological and evolutionary life-sciences. The introduction of the problem of life-sciences leads in turn to psychology as the turning point and connection between the natural and the human sciences.

Here Seebohm speaks of the controversies that raged at the turn of the twentieth century over the nature of psychology. It was emerging as an experimental discipline, but this was countered by Dilthey's project for a humanistic psychology and Husserl's eventual development of phenomenological psychology. But this brings us to what Seebohm (following Dilthey) calls the "systematic," i.e., the non-historical human sciences. These include, in addition to psychology, sociology, economics, jurisprudence, and political science. Since these disciplines deal with the present world they do not have the problem of interpreting the "fixed life expressions" of past and alien lifeworlds. Thus, to a large extent they share the lifeworld of their subjects, i.e., the persons and groups whose activities and behaviors they study. One twist in the understanding of some of these disciplines is that they are often pursued more as practical than theoretical sciences, even though they rely on theoretical premises.

(3) Now I want to turn my attention to the broader issues of historicism and the paradox of subjectivity, which occupy Seebohm at the beginning and the end of his book. When I say broader issues, I mean that these questions concern not just the system of the sciences, in the sense of a systematic epistemology and phenomenology of the sciences, but rather the status of phenomenology itself. Ultimately these issues do have a bearing on the "system of the sciences" in the sense that they concern the overall interpretation of such a system, but this is because the system is proposed by Seebohm as a phenomenological and not a metaphysical system.

I mentioned at the beginning that the structure of *Ideas II* represents Husserl's early conception of what a system of the sciences might be. I pointed out that it was very conventional in the sense that it starts with the natural sciences and proceeds to build the psychological and human sciences upon them. However, it is well known that by the time he reaches the third part, at least in the version edited by Stein and Landgrebe, important changes are afoot. I want to turn briefly to these changes.

The third part begins with an extended reflection on the concept of "attitude." Constitution always presupposes an attitude toward the world and the objects in it, and the attitude tacitly presupposed in the discussion of the constitution of nature has been that of the natural sciences. Even the discussion of "animal nature" had been oriented toward the idea of a scientific psychology. Animals and humans are perceived as objects in the natural world, but when we take them as "ensouled" beings we are aware of sensitive and psychic properties and capacities which constitute a "*Realitätsüberschuß*," a surplus over the physical (Husserl 1952, 176). Here Husserl uses the word "*Introjektion*"—though he also notes that this term can be "easily misunderstood."

In his transition to the "*geistige Welt*" Husserl notes that one great merit of the phenomenological reduction is that it can free us from the limitations imposed by "the natural attitude or any relative attitude," limitations that are not noticed by "natural man, [*natürlicher Mensch*] and in particular the scientist of nature [*Naturforscher*]" (179). Terminologically, Husserl in this passage is still identifying the "natural attitude" with that of the natural scientist. But then he suggests a distinction between "*natürlich*" and "*naturalistisch*," reserving the latter for the attitude of the natural scientist, and pointing out that the attitude toward persons, communities and the social world is "*sehr natürlich*" (180). What he now calls the "personalistic attitude, which we are always in when we live together, speak to one another," etc., is "a thoroughly natural and not an artificial attitude that has to be attained and maintained by special means" (183).

Having articulated the differences between the naturalistic and the personalistic attitudes, Husserl describes how the scientist, like everyone else, lives in the personalistic attitude, but for purposes of scientific inquiry assumes the naturalistic attitude. He now arrives at the following, rather striking conclusion:

> Looked at more closely, it will even become clear that we are not at all dealing with two attitudes, possessing equal rights and equal status, two equal and mutually interpenetrating apperceptions, but rather that the naturalistic attitude is subordinate to the personalistic and has acquired a certain independence through a kind of abstraction, or rather through a sort of self-forgetfulness of the personal ego, at the same time unjustifiably absolutizing its world, [i.e.] nature. (183–184)

We can now see that the "world of the natural attitude," sought by Husserl in the early pages of *Ideas I*, has been completely transformed. Although in some sense, in that text Husserl wanted to capture the naïve and unsophisticated view of the world; in fact, the natural attitude was correlated with the scientific view of nature and represented the view of the natural scientist, or at the very least the incipient or aspiring natural scientist. Now the natural attitude is that of the non-scientist, or even the scientist in his or her off-hours, and the *world* of the natural attitude is the world of persons, of motivations, of social relations, of communities, and of "spiritual" entities and predicates. Here natural science is construed as a social enterprise, and nature makes a return appearance, not as something absolute, but as "something constituted in the intersubjective band of persons" that make up the scientific community (210).

While Husserl seems here to be asserting the priority of the spiritual over the natural world, he nevertheless ends this section of *Ideas II* wondering if we are left with a "two-worlds" theory, or a theory of "fundamentally different 'regions.' Is it really a matter of two different worlds, 'nature' on one side, the world of the spirit on the other, separated by cardinal ontological distinctions?" (210). The two worlds need not be unrelated: the spiritual from the naturalistic standpoint can be seen as an "annex" of the physical body; and the natural, as we have seen, has its place within the spiritual world. But Husserl seems dissatisfied with this outcome for two reasons: first, the seeming equality of the two sides and, second, the fact that the *world* seems to have lost its unity, sundering itself into the kind of metaphysical dualism that resembles that of Descartes. At the end of *Ideas II*, Husserl seems to conclude that the one side of this dualism swallows the other up: the last chapter bears the title "The Ontological Priority of the Spiritual World over the Naturalistic" and concludes with the section called "relativity of nature, absoluteness of spirit." This, in my view, would be to replace a metaphysical dualism with a metaphysical monism—actually a metaphysical or ontological idealism.

But this solution to the unity of the world seems to operate at a level different from that of the "world of the natural attitude." The "phenomenological investigations on constitution" that make up the three parts of *Ideas II* are very much geared to the emergence of the sciences: the natural and psychological sciences in the first two parts, and the *Geisteswissenschaften* in the third. Each of these corresponds to a different "region of reality" addressed by each of these sciences. Husserl seems to presuppose these regions as fully developed scientifically when he raises the issue of their equality or non-equality, and then proposes that one swallows up the others.

But in the lectures on phenomenological psychology of 1925, Husserl looks elsewhere for the unity of the world and the relation among the regions. In this text, in which Husserl also returns to Avenarius's expression *der natürliche Weltbegriff*, he writes the following:

> As scientific themes, nature and mind do not exist beforehand; rather, they are formed only within a theoretical interest and in the theoretical work directed by it, upon the underlying stratum of a natural, prescientific experience [*Erfahrung*]. Here they appear in an originally intuitable intermingling and togetherness; it is necessary to begin with this concretely intuitable unity of the prescientific experiential world and then to elucidate what theoretical interests and directions of thought it predeliniates, as well as how nature and mind can become unitary universal themes, always inseparably related to each other, in it. (1962, 55)

Thus, the unity of the three regions is to be found not at the scientific level but in the pre-scientific world of the natural attitude, a conception that presages that of the lifeworld in the *Crisis*. In other words, the world is unified prior to and independently of its being theoretically divided into regions, not after it has been so divided.

The view expressed in this passage from "Phenomenological Psychology," as further developed in the *Crisis*, completes the transformation of Husserl's concept of world, and goes beyond even the important changes found in *Ideas II*. There the attitude underlying the human sciences is found to be more "natural" than that of the natural sciences, and Husserl expresses the view that the spiritual world in a sense encompasses the natural world and renders it secondary and dependent. But the lifeworld of the *Crisis*, and the natural attitude (or natural world-life, as Husserl sometimes calls it) that corresponds to it, is prior to all regions, prior to all scientific or even proto-scientific attitudes, prior even to all practical attitudes that divide up the world for some purpose or other. As Husserl says in the *Crisis*, the lifeworld is

> that which is *taken for granted,* which is presupposed by all thinking, all activity of life with all its ends and accomplishments… . Before all such accomplishments there has always already been a universal accomplishment, presupposed by all human practice and all scientific and prescientific life. The latter have the spiritual acquisitions of this universal accomplishment as their constant substratum, and all their own acquisitions are destined to flow into it. (Husserl 1954, 115)

And in one of the *Beilagen* he writes:

> The life-world is that world that is constantly pregiven, valid constantly and in advance as existing, but not valid because of some purpose of investigation, according to some universal end. Every end presupposes it; even the universal end of knowing it in scientific truth presupposes it… Now though we must [further] make evident the fact that the life-world itself is a "structure" [*Gebilde*], it is nevertheless not a "purposeful structure…" (461)

It is in this context, of course, at the end of the long section of the *Crisis* devoted to the "way into transcendental phenomenology" from the pre-given lifeworld, that Husserl articulates the "paradox of human subjectivity: being a subject for the world and at the same time being an object for the world" (182). I especially like Seebohm's take on this passage. He points out that Husserl's way of articulating this paradox emphasizes the ontological: *being* a subject [*Subjektsein*] versus *being* an object [*Objektsein*] (Seebohm 2015, 39). Given the radical ontological disparity between these two descriptions (being a subject and being an object) how can one entity be both? Hence the paradox. This makes it sound as if the human subject is being located within two ontological regions, which might be thought to correspond to the two attitudes we have discussed, the naturalistic and the personalistic. And since Husserl articulates this as an ontological paradox, it must have an ontological solution. This is what we find at the end of *Ideas II*: remember the chapter title we quoted from that text: "The *Ontological* Priority of the Spiritual World over the Naturalistic" (Husserl 1952, 281). Section 64, the very last in this text, is entitled "relativity of nature, absoluteness of spirit" (297). So, we arrive at the Absolute Spirit, a familiar figure in the history of German philosophy.

As Seebohm points out, the solution to the paradox in the *Crisis* is similarly ontological: the "Ur-ich" or primal I, the absolute subject lying behind all worldly subjects. Thus, Husserl succumbs to the temptation which seems to lie in the German philosophical DNA, the desire for "*Letztbegründung*" or ultimate ontological—or metaphysical—foundation. Here Husserl was influenced in his last years by Fink, who in turn, and ironically, was influenced by Heidegger. Seebohm calls this "Husserl's and Fink's turn to a speculative metaphysical interpretation of the transcendental Ego" (2015, 388) and "Fink's quasi-Fichtean solution" (387).

Seebohm proposes, however, that we look at the difference between the two descriptions of human subjectivity not as an ontological difference but as an epistemic or epistemological difference. This difference indeed corresponds to two different attitudes, not to the difference between the naturalistic and the personalistic attitudes, but rather to the difference between the *natural* (mundane) attitude and the phenomenological attitude. I would prefer to say that this difference can be interpreted as a methodological rather than an ontological or metaphysical difference. Phenomenology proposes that we consider the human being as subject for the world, while the natural attitude places the human being, like everything else, within the world. This then gives us not two different things or entities, but two different ways of looking at the *same* thing. On this interpretation, As Seebohm tells us several times, *there is no paradox* (2015, 39, 40). And hence no need for a solution.

But this requires that we treat phenomenology strictly as a method which is metaphysically neutral. As Seebohm points out, this interpretation was shared by those who rejected the Finkian metaphysicization, including the phenomenologists of the New School: Schutz, Cairns, Kaufmann, and Gurwitsch, together with Seebohm's own teacher Gerhard Funke.

I consider myself a late-generation member of this school, and I have advanced a similar interpretation of the paradox of subjectivity. I don't think we can say that there is no paradox, as Seebohm does. I think we cannot avoid the opposed descriptions of the human subject, so we cannot make the paradox disappear. But we can't solve it either, so we just have to live with it.

References

Husserl, Edmund. 1952. *Ideen zu einer reinen Phänomenologie. Zweites Buch: Phänomenologische Untersuchungen zur Konstitution*, Husserliana IV. Ed. Marly Biemel. The Hague: Martinus Nijhoff.

———. 1954. *Die Krisis der europäischen Wissenschaften*, Husserliana VI. Ed. Walter Biemel. The Hague: Martinus Nijhoff.

———. 1962. *Phänomenologische Psychologie*, Husserliana IX. Ed. Walter Biemel. Haag: Martinus Nijhoff.

Seebohm, Thomas M. 2015. *History as a science and the system of the sciences: Phenomenological investigations*. New York: Springer.

David Carr is Charles Howard Candler Professor Emeritus at Emory University (Atlanta) and Adjunct Professor of Philosophy at the New School for Social Research, New York City. His areas of research are twentieth-century phenomenology, especially Husserl, theory of historical narrative, and philosophy of history. He is the author of *Phenomenology and the Problem of History* (1974, reissued in 2009); *Time, Narrative, and History* (1986); *Interpreting Husserl* (1987); *The Paradox of Subjectivity* (1999); and *Experience and History* (2014).

Chapter 2
Seebohm and Husserl on the Humanities

Thomas Nenon

Abstract This essay discusses Husserl's brief and Seebohm's much more extensive comments on the grounding of several disciplines commonly described under the heading of "the humanities" as sciences in his *History as a Science and the System of the Sciences*. It asks not just what Husserl and Seebohm say about the grounding of these disciplines as empirical sciences, but also whether we can understand what they have to say by applying what they have to say to the actual practice of these disciplines, thereby extending and refining some of the rather general claims they make about the way they should and do proceed as genuine sciences in the sense indicated above. It does not attempt to do this with all of the humanities but with a few paradigmatic examples, all of which, as academic disciplines, typically take *texts* and the matters these texts are directed to as the objects of their studies. It thereby becomes clear in the summary of Husserl's and Seebohm's positions on the grounding of the *Geisteswissenschaften* (Husserl's term, not Seebohm's) why the notion of interpretation is so important, in particular for literary studies and history, but not so much for applied disciplines such as composition studies or foreign language instruction that are part of humanities departments but do not fit the classic model of the *Geisteswissenschaften* as theoretical, interpretive disciplines.

This essay will discuss Husserl's brief and Seebohm's much more extensive comments on the grounding of several disciplines commonly described under the heading of "the humanities" as sciences in his *History as a Science and the System of the Sciences* (2015). Of course, in the English-speaking world, as opposed to the German-speaking world, practitioners of these disciplines usually do not refer to the work they do as science, but most practitioners do make claims that they believe fit

T. Nenon (✉)
Department of Philosophy, University of Memphis, Memphis, TN, USA
e-mail: tnenon@memphis.edu

© Springer Nature Switzerland AG 2020

T. Nenon (ed.), *Thomas Seebohm on the Foundations of the Sciences*,
Contributions to Phenomenology 105,
https://doi.org/10.1007/978-3-030-23661-8_2

Seebohm's (and Husserl's) definitions of science as a set of claims that can be verified and justified according to some established measures. In the German-speaking world where these disciplines would still be commonly referred to as *Geistes*- or *Kulturwissenschaften*, the idea of them as sciences does not seem strange at all. What would seem strange is the assumption in English-speaking countries that only the social sciences and the natural sciences would or could be real sciences. Seebohm also makes a convincing case that, if the *Geisteswissenschaften* are sciences, then they are *empirical sciences* since the opposite for him is *formal* sciences, which they certainly are not. Even in the German-speaking world, most practitioners of these disciplines would not think of themselves as doing empirical work since that term is normally associated with lab experiments or statistical methods instead of interpretive work, but if we accept Seebohm's taxonomy, then interpretive work too is certainly empirical.

Moreover, I would like not just to discuss what Husserl and Seebohm say about the grounding of these disciplines as empirical sciences, but also see if we can understand what they have to say by applying what they have to say to the actual practice of these disciplines, thereby extending and refining some of the rather general claims they make about the way they should and do proceed as genuine sciences in the sense indicated above. I will not attempt to do this with all of the humanities but with a few paradigmatic examples, all of which, as academic disciplines, typically take *texts* and the matters these texts are directed to as the objects of their studies. So I will talk about English, foreign languages, history, and philosophy, but I will not discuss musicology, music history, or art history, for example. Of course, not all of the areas typically included in departments of English, foreign languages, history, and philosophy actually are or claim to be sciences in Seebohm's sense. For instance, foreign language instruction, English as a second language, composition studies, and technical writing aim not at valid interpretations even when they have to do with books and texts, but at imparting skills, and to the extent that they not only teach and practice skills, but also investigate the best ways to impart those skills, they employ statistical methods closer to those of the social sciences than the other areas of the humanities, and the latter two areas concentrate more on how to produce good texts than how to interpret texts composed by others. It will become clear in the summary of Husserl's and Seebohm's positions on the grounding of the *Geisteswissenschaften* (Husserl's term, not Seebohm's) why the notion of interpretation is so important, but I will return to that topic a little later.

Seebohm sees his own work as extending and refining Husserl's work. However, he also notes that Husserl's own contributions to the grounding of even the formal and the natural sciences remains rather general and requires some significant refinements, and that Husserl has even less to say about the specific methods of and differences between various sciences that he lumps together under the heading of the *Geisteswissenschaften*. Here Seebohm is certainly correct. It is perhaps no surprise that Husserl, originally coming from mathematics as his home discipline, had more and much more specific things to say about the grounding of formal sciences, beginning as he did with the philosophy of arithmetic in his first larger work and then turning to questions about the foundations of formal disciplines including logic in his famous refutation of psychologism in the *Prolegomena to the Logical Investigations*, and devoting one of the few major works published in his lifetime, *Formal and Transcendental Logic*, to the grounding of formal logic. By contrast, the *Logical Investigations* themselves, according to Seebohm's reading, which I think is correct, are meant to be contributions to the grounds of any science at all, including for instance, analyses of meaning, the logic of parts and wholes, fundamental grammar, intentionality, and truth, that provide a contribution to a formal ontology underlying any specific ontologies of various disciplines related either to ideal or real objects. The third major work published during his lifetime, the *Ideas I*, was devoted to the methodology of phenomenological philosophy, intended as the general method that would guide the development of the regional ontologies that are at the foundation of different scientific domains. The one remaining major work published during his lifetime, the *Crisis*, is where he actually does offer extended analyses of the foundation of some of the material sciences, most particularly the origin and foundation of modern mechanistic sciences that apply some of the formal methods of mathematics to the study of nature in a highly abstractive way that fundamentally changes what we mean by "nature." As a historical-genetic study, the *Crisis* shows how the formal disciplines of mathematics and modern natural science have their origins in a much more concrete reality, namely the lifeworld, that cannot be reduced to nature in this new sense as the correlate of the natural sciences and cannot be captured by them. One of the implications of this would be that the methods of the natural sciences would not be appropriate to ground other sciences that deal with different kinds of objects that we encounter in the lifeworld other than objects taken in this very abstract and limited sense as natural objects. In the *Crisis* itself, Husserl says almost nothing about how this would work except in this general contrastive way, namely that the ontologies of lifeworldly objects would be different and that the method for approaching them would therefore need to be different as well.

As we learn in several of the other chapters in Seebohm's *History as a Science and the System of the Sciences*, one most important source for insights into how Husserl conceived of these regional ontologies is his *Ideas II*, where he explicitly and fairly extensively describes the differences in attitudes that define the regions first of nature, then of souls [*Seelen*], and finally of *Geist*. The region of *Geist* is the correlate of the personalistic attitude. The personalistic attitude (as opposed to the naturalistic attitude of the natural sciences) is our natural attitude in the sense that it is the attitude in which we live our daily lives and in which we recognize other human beings as persons, live in communities with social identities, and interact with all sorts of things as use-objects. This everyday world, the *Umwelt*, is what Husserl will subsequently examine and describe as the lifeworld. In another, much longer paper, I have tried to lay out the development of Husserl's conception of the lifeworld and its basic structures based on the extensive research manuscripts on the topic of the lifeworld in Volume XXXIX of the *Husserliana* (see Nenon 2014). This volume does indeed present a rich trove of observations and insights that would be relevant to the human and social sciences, but it remains true both that everything he writes in these research manuscripts is consistent with the general direction of the third section of the *Ideas II*, and that in the nearly 800 pages in Volume XXXIX, one still cannot find the kind of specific studies into the foundation of the human or social sciences that Seebohm undertakes in his work on *History as a Science and the System of the Sciences*. It seems that this remained the kind of work that Husserl intended to farm out to one of his students in the way that he assigned one of them to the phenomenology of mathematics, another the phenomenology of religion, etc. All the more laudable then, that Seebohm takes on at least the beginnings of all of these assignments in one systematic work.

Before we turn to Seebohm, though, it is important to recall a few of the very general things we do learn from Husserl about the basic differences between the natural sciences and the sciences of *Geist*, and a couple of the fundamental insights Husserl's phenomenology achieves that make the kind of work that Seebohm undertakes possible at all. I mentioned above that in the attitude in which we live our daily lives, we encounter the objects around us in terms of their significance for us, what they are good for, how they can serve or threaten our ends, how they fit in with what we value. It is a world in which we encounter other human beings and ourselves as persons and, for Husserl, that means as embodied agents who are guided by beliefs, values, and desires. It is a social world, in which we also recognize ourselves and others in terms of certain social identities and recognize and know how to operate in different kinds of social organizations. It is a historical world, in which we and those social organizations have been shaped by our individual and common experiences and in which we grasp ourselves, others, and other things prospectively in terms of how they relate to the means and ends we have. Moreover, the means and ends will have something to do with how we are embodied, and even the way we perceive and encounter things will depend on the way we are embodied, individually and socially. It is a world shaped by what Husserl calls in the research manuscript "relativities" to our specific forms of embodiment, our own beliefs, values, and ends that are socially and historically conditioned, and our specific locations in concrete time and space. Moreover, it is also a world in which we, as persons, think of ourselves as capable of reflection and taking responsibility for our beliefs, actions, and desires, that is, as "free" in this sense (see Nenon 2002). By contrast, the project of the natural sciences is to obtain "objectivity" through abstraction from all such relativities, by attempting to reduce all of the genuine predicates of objects to predicates that are quantifiable and calculable, to thereby achieve testable truths that are not relative to any society or age, nor even relative to human perception and the concepts relative to it—the so-called secondary qualities of perceptual objects that are as much a result of specifically human and variable forms of embodiment as the "true" qualities of objects such as the wavelengths of the photons they emit or their mass.[1]

[1] I do not know whether Seebohm recognizes that one difference between the physicalist sciences and the life sciences is that the latter are not nearly as intent on avoiding secondary-quality descriptions of the objects of their work as the former tend to be. Most biologists have no problem describing their objects in terms of colors or smells instead of wavelengths or grams of something per square meter when describing larger plants and animals. Of course, at the increasingly dominant

Moreover, their findings are expressed in terms of statistical probabilities or even exceptionless laws with a deterministic mode of explanation. In a positive sense then, we can conclude that the *Geisteswissenschaften* will not be subject to these limitations, that they will somehow be able to study and draw valid conclusions about such things as the values and ends that persons have, about their social organizations, about their histories as individuals and as groups, about various kind of objects that are meaningful to them, and how they are meaningful to them.

The sciences that are based on our everyday world of experience in the lifeworld are not limited to the kinds predicates that are acceptable to the natural sciences; it is important to emphasize what is perhaps Husserl's most important contribution to the grounding of the human and social sciences, namely that reality is not one kind of thing, but that there are various regions of objects, each of them characterized by different kinds of predicates that can obtain for them, and each of which has its own kind of intuition or experience of verification that can reliably tell us which of these different kinds of objects exist and what they are like. An attitude is a way of looking at things and in which different kinds of things can show up for us as genuine, and they can exhibit different kinds of properties that we have ways of ascertaining they really do have or not. The project of phenomenology is to reflect on the different kinds of experiences we have in which different kinds of objects show themselves to us and to see how the different kinds of experiences and different kinds of objects are related to each other in "foundational" relationships. It is against this backdrop that the project of a grounding of the human and social sciences can make sense as a study of the refinements and systematic development of methods of investigation and verification—one that begins with our everyday experiences and everyday process of confirmation and disconfirmation of beliefs, values, and practices and shows how the various human and social sciences emerge out of them just as the natural sciences do as well, but without the specific abstractions that make the natural sciences possible. The first and key step is to overcome the false belief, the reductionist tendency, according to which the only things there really are and the only properties those things really have are the ones that are accessible to natural science and therefore that it is the only true science and only its methods are valid. Seebohm explicitly draws upon this phenomenological project in the preface and the first few pages of this work (2015, v–vii, 1–2).

This does not entail that the methods of the natural sciences are irrelevant for the human sciences. Husserl's notion of "foundation" allows for a method that builds upon, but is not reducible to, the methods of lower-level phenomena. Since persons are higher-level entities that are founded upon, but not reducible to the physical aspects of their existence as embodied agents, then facts arrived at through the methods of the natural sciences, for instance, facts about human biology and human brain psychology, are in a sense presupposed in any interpretation of their behavior,

---

genomic level, the descriptions are all in terms of chemistry and thus usually described in physicalist predicates. To the extent that many biologists are increasingly attempting to give genomic accounts of almost everything, this would undermine some of that difference, but that project is still in its infancy.

even behaviors we classify as motivated actions, and may play a role in how we interpret those actions. Similarly, facts about geography, weather, and other physical circumstances will play a role in the understanding of the constitution and actions of even higher-level personalities such as communities or states, but are far from sufficient for the kind of understanding that the human sciences aim at and often achieve. Moreover, these natural phenomena that are relevant for the human sciences will normally not require some specific competence in the natural sciences or typically employ technical terms unique to the natural sciences, but rather everyday common understandings of the phenomena described there.

It is relevant to note that Husserl did not rule out using practical predicates for natural science either, and indeed we know that scientists will refer to the kinds of instruments they use to obtain their results using the practical language of tools, which would not surprise him based on his view that the natural sciences still depend on the lifeworld. His more important and significant insight here is the emphasis in his later work that the natural sciences are grounded in the lifeworld as a specific historically developed and social practice with specific aims. This is consistent with Seebohm's claims but inconsistent with many scientists' self-understanding and the ontologies and epistemologies of philosophers and others who want to take the abstractive methods of the natural scientific abstraction as their exclusive model for what exists and what is true. In fact, Seebohm's entire project of showing how these very different sciences have different equally valid methods is meant as a repudiation of that assumption.

For Seebohm, what is crucial about human sciences like philology and the historical sciences is that their primary concern is interpretations or "understanding" or that at the very least they presuppose interpretations even when their interest is a causal explanation of past events, as it often is in history as a human science, for example (vi). This is because the kinds of events that history seeks to explain are manifestations of human "cultural activities," i.e., actions. Hence, he recognizes understanding as a basic category for the human sciences in spite of his rejection of a clear dichotomy between explaining and understanding as basis for distinguishing the natural from the human sciences.

In this regard, Seebohm finds himself basically in line with a tradition that he notes includes Dilthey and Husserl. However, he claims that they had failed to distinguish clearly enough very different kinds of understanding that are at work in our daily lives and in the human sciences. Seebohm makes a distinction between first-order elementary understanding, second-order elementary understanding, and both of these from what he calls higher-order understanding:

Underlying both of these levels, however, is a more basic level that he calls "animalic understanding." Having described the genetic origins of object-awareness through passive synthesis, Seebohm describes how we become aware of ourselves and others as living bodies when: "The life expressions of the other living body, trigger as actions of the other living body immediate reactions of one's own living body and vice versa," hence limiting it to the expressions and not the intersubjectively given states of affairs (47). He subsequently expands this description by including not only the immediate understanding of the life expressions of others as animalic understanding, i.e., "life expressions as indicators of the intentions and feeling of others" but also the "reference of life expressions to intersubjectively given states of affairs" (49). Either way, it is clear that these are pre-reflexive forms of understanding that necessarily involve an awareness of others as others. "Animalic understanding of animalic life expressions is the genetic foundation of elementary understanding and higher understanding" (49).

This most basic level of understanding provides the foundation for the others:

> Elementary understanding will be characterized as the understanding at work in the interactions of practical life and higher understanding as the contemplative theoretical understanding of the lifeworld, of the human condition in religion, art, philosophy, and last but not least, the sciences. First-order understanding is the understanding of intentional objects given in passive and active cognitive syntheses, and second-order] understanding the understanding of the first-order understanding of Others. (5–6)

At all of these levels, the objects are somehow related to the life expressions of living beings. They are things that we understand, fail to understand, or misunderstand in our daily lives. Seebohm list several kinds of them. Obviously, it must include the words and gestures of others, but it also includes, at the elementary level, natural objects that serve as tools for our needs and would also include the artifacts produced by them that reflect the needs, practices, and values of the persons who use them. These are the kinds of things that Husserl describes in the third section of the *Ideas II*, where he describes the personalistic attitude as use-objects; or what Heidegger in *Being and Time* calls the "ready-to-hand [*Zuhandenes*]." These kinds of thing are often or even usually produced or shaped by human beings, but not necessarily so, since we can recognize, classify, and use or consume naturally produced things as use-objects or cultural objects as well. It would also include the artifacts that the tools are used to create. In most cases, though, these tools and other artifacts refer back to their producers as their "authors" (49).

Higher-level understanding, Seebohm says, goes beyond the fulfillment of immediate needs and interests:

> The first and main task of *higher* understanding is "to make sense," to understand and explain the blind forces, human forces included, behind the natural environment as a correlate of elementary understanding.
>
> What is understood in higher understanding is the lifeworld as a whole and its essential aspects. Higher understanding is contemplative. (51)

As examples of the manifestations of such high-order understanding, Seebohm lists religious cults, myths, art works, poetry, philosophical reflections, and finally scientific treatises and articles as ways of making sense of the world as a whole (51–52). All of these various kinds of things can thereby also become the objects of the human sciences.

Another important distinction that Seebohm does make is the difference between "immediate and fixed life expressions":

> Immediate life expressions … are given in the actual present and its immediate past. Fixed life expressions can be given intersubjectively in the actual present, but they can also be given as life expressions of the authors of a distant past beyond the scope of subjective memories and as the same again in the future. (49)

The distinction is crucial because it is the latter class of objects, namely the fixed expressions, that are the objects of the human sciences of philology and history, among others. Moreover, it is precisely because they can become objects of understanding in a subsequent context, a lifeworld or horizon, that is very different from the context in which they arose that gives rise to the problems that Seebohm discusses as the first and second canons of hermeneutics. First-order higher understanding is directed to the events, objects, actions, and life expressions of the current lifeworld from the cultural horizon of those who experience them. Second-order higher understanding, by contrast, is directed to the first-order fixed life expressions, especially those fixed in written form (56). Hence, the kind of understanding that interests Seebohm in his description of the system of the sciences is second-order understanding of fixed first-order understanding, and, in particular, the kind of understanding sought in philology and in history as sciences.

Turning his attention to philology, Seebohm describe how modern literary philology can be traced back to the attempt of previous cultures to understand sacred texts across time and how to be able to continue to discover their "original truth" (57). It should be noted that, at this stage, the question of their truth is also an eminently practical one, namely, whether and how we should follow what we take to be its dictates. Reflections on the appropriate ways to do this gives rise to the more general question of how to understand and adjudicate the truth claims of all sorts of sacred and then later legal and political texts that pave the way for what will later emerge as philology:

> In the early phases of the development of a literary culture, the trust in the truth and unity of the oral and/or written tradition is a necessary condition for the stability of the social structures of archaic cultures. The trust in the tradition and the need to apply it again in the present and its future horizon vanishes step by step in the development of a literary tradition with different meta-genres, with the increase in the amount of rejections of parts of the old tradition, and with their replacement by new "truths." The development of, e.g., an independent legal literature and/or independent philosophical reflections about right and wrong, good and evil, and last but not least nature is the generative foundation for partial rejections of the contents of the religious literature, and finally even of religion itself. Poetry and art are freed from the fetters of their functions in religious services and cults can secularize certain religious contents and give preference to worldly perspectives. (58)

Seebohm describes how modern philology emerged as classical traditions of philology were rediscovered and used by the humanists of the Renaissance "in their interpretations of the texts of the classical literary tradition. The tool was then also used by Protestant theologians. The final consequence was the rejection of the medieval tradition as a tradition of the 'Dark Ages'" (67). Philology became the accepted mode for the interpretation of traditional texts and archeology for the interpretation of buildings and monuments. The immediate objects of study now shifted to interpretations of one's own still presumably valid tradition to objects from traditions temporally or geographically distant, and hence different from one's own:

> The experience of what is foreign and has even been rejected and forgotten in one's own tradition, together with the experience of highly developed foreign cultural lifeworlds in the sixteenth, seventeenth, and eighteenth centuries was the presupposition for the genesis of the historical human disciplines of philology, archaeology, and history as sciences. (68)

This distance was also accompanied by a suspension in belief in their truth claims to the extent that the question of their possible applications or rejections was no longer of interest in their fields as such: "The task was to interpret the sources of past and present foreign cultures and their historical development in their own context" (68). This is part of the reason why Seebohm stresses that philology as a science is not "literary criticism" (which in his view is not a science but based on one, namely philology) (2). The point of these new disciplines was now simply to understand in the sense of placing the objects of their study into the appropriate historical and cultural contexts in which they were created.

Seebohm stresses that both philology and history as sciences are forms of higher understanding. However, they differ in that philology as a stand-alone discipline, whether as classical philology or the more recent philological disciplines of modern languages and literatures, has as its primary objects literary artworks, that is, manifestations of first-order higher understanding (101). Of course, philology as a method rather than as a stand-alone discipline is still relevant for legal and sacred texts. By contrast, the objects of history are much more all-encompassing, including not just texts but all kinds of first-order life expressions and even natural events to the extent that they had an effect on human actions and institutions. At the same time, many of the traces that history studies are texts, and often texts in languages that are not our own or from times prior to significant changes in the language we speak today, so philology is often a pre-condition for historical research.

Seebohm's descriptions of the human sciences fits most closely with the work performed in departments of history in the United States, in which explanations of actions, events, and institutions in terms of their historical and social contexts remain at the center of their work. This remains true even of various version of critical histories, which focus on and seek to uncover strategies of oppression or repression at work that traditional accounts have neglected or intentionally covered up. The goal is still to provide an account that is closer to the "true" nature of what was actually at work instead of what the dominant regime claimed.

In literature departments, whether in English or foreign languages, the American tradition of interpretation was different from Germany's. The notion of literary studies as a *Geisteswissenschaft* in Germany tended to reinforce the role of philology as the core enterprise of literary studies until relatively recently. In the United States, literary studies until recently had often been associated with notions of "universal truths" expressed in great works of art that in a sense serve as secular versions of holy scriptures. Ironically, it has been critical literary studies that can be seen to be versions of traditional hermeneutics, as they also place literary works into critical contexts that they exhibit as part of larger social patterns.

This is also the case for foreign languages, where philology is part of what many of them do as a research area, philology not as the study of Literature with a capital L, but of grammatical and lexical structures of a language as such, where the texts themselves are more the means than the end at first. However, in America these departments typically include literary studies and linguistics, both versions of philology, but also simple language instruction as a large part of what they do, as well as cultural studies; just like English departments usually include composition studies, creative writing, and often English as a second language studies in addition to the traditional philological disciplines.

Other interpretive studies that take objects other than texts as their objects, namely art history and music history, are often included in the humanities, depending in some cases on where they are located. Most often they will be subordinate parts of larger departments in which music or art theory and the practice and production of art and music are the main interests of those departments, so they end up being included under the heading of the fine arts even though their methods are closer to those of literary studies.

Of course, in both Germany and the United States, philosophy itself is pursued not just as a historical study of texts, i.e., philology, but as a high-order enterprise of understanding itself, which attempts not only to explain and contextualize the texts of historical and contemporary others, but to critically evaluate them as adequate or inadequate and to advance such higher-order understandings on their own.

Overall, then, one observation about Seebohm's work and the humanities as understood in the United States, and perhaps in English-speaking countries, is that Seebohm's insights into the history and nature of philology and history are indeed directly relevant for departments of German and foreign literatures and history in German universities, and they apply to some core parts of the work pursued in humanities departments in America, but the practical nature of American higher education has rendered theoretical interpretive studies as just one part of what is understood by "the humanities" in that tradition.

References

Nenon, Thomas. 2002. Freedom, responsibility, and self-awareness in Husserl. In *Neues Jahrbuch für Phänomenologie und phänomenologische Philosophie*: 1–21.

———. 2014. Intersubjectivity, interculturality, and realities in Husserl's research manuscripts on the lifeworld (Hua XXXIX). In *The phenomenology of embodied subjectivity*, ed. Dermot Moran. New York: Springer.

Seebohm, Thomas M. 2015. *History as a science and the system of the sciences: Phenomenological investigations*. New York: Springer.

Thomas Nenon (Ph.D. 1983, University of Freiburg, Germany) is a Professor of Philosophy and Provost at the University of Memphis. He worked as an editor at the Husserl Archives and instructor at the University of Freiburg before coming to the University of Memphis. His teaching and research interests include Husserl, Heidegger, Kant and German idealism, hermeneutics, and the philosophy of the social sciences. He has published numerous articles in those areas as well as the book *Objektivität und endliche Erkenntnis* (Freiburg: Alber 1986) and is co-editor (along with Hans Rainer Sepp) of volumes XXV and XXVII, as well as editor of several collections of essays. He has served as review editor for *Husserl Studies*, as a member of the Executive Committee of the Society for Phenomenology and Existential Philosophy, as Director of the Center for the Humanities at Memphis, and as President of the Center for Advanced Research in Phenomenology. His current research interests include Husserl's ethics and theory of values as well as Kant's and Hegel's practical philosophy.

Chapter 3
History, the Sciences, and Disinterested Observers: A Dialogue Between Alfred Schutz and Thomas Seebohm

Michael D. Barber

Abstract Thomas Seebohm does not recognize sufficiently that Alfred Schutz would have been open to the contributions the natural sciences could make to the cultural sciences and to the use of predictive methodologies in the cultural sciences. Further, in contrast to Seebohm, Schutz takes into account the minimal ontological differences underpinning the natural science/cultural science divide. As far as disinterested observers are concerned, Schutz was clear about the differences between past and present and the need for correlative epistemological adjustments. In addition, Schutz shows the importance of the relevances of the cultural scientist, such as the positive interest in discovering the historical meaning of past actors, for achieving social scientific objectivity, and he is attuned to the importance of the community of scientists for determining objectivity. Consequently, his concern for cultural scientific objectivity is less ascetical in character than Seebohm's. Further differences between Schutz and Seebohm appear in their understanding of because motives, the relationship between the natural attitude and the life-world, and the use of eidetic methodology outside of the parameters of the phenomenological reduction.

This paper will examine history's place among the cultural sciences by developing a dialogue between Thomas Seebohm and Alfred Schutz—a dialogue that is at least inchoate in the former's book *History as a Science and the System of the Sciences: Phenomenological Investigations*. Since history's unique relationship to the natural sciences permits it to serve as a "mediator" (Seebohm 2015, 95, 263) between the natural sciences and the cultural sciences, according to Seebohm, it is important to discuss the relationship between the natural and cultural sciences and history's place in that relationship. A second, related issue has to do with the possibility of being "a disinterested observer" in history and the other cultural sciences. That discussion must take account of history's distinctiveness with regard to the other cultural sciences since the question of the disinterested observer arises in that setting

M. D. Barber (✉)
Saint Louis University, St. Louis, MO, USA
e-mail: michael.barber@slu.edu

© Springer Nature Switzerland AG 2020

T. Nenon (ed.), *Thomas Seebohm on the Foundations of the Sciences*, Contributions to Phenomenology 105,
https://doi.org/10.1007/978-3-030-23661-8_3

and since Seebohm specifically criticizes Schutz's account of that relationship. Finally, there are a series of minor areas in which Seebohm seems at odds with Schutz and Husserl and to which some discussion deserves to be given.

3.1 Seebohm on the History and the Cultural/Natural Science Divide

One of the great merits of Seebohm's book is that it illustrates the many, diverse ways in which history inescapably relies upon the natural sciences as constitutive of its own endeavors. Seebohm thereby undermines the Rickertian/Windlebandian separation of the natural sciences from the cultural sciences because the former are nomothetic and causal in orientation and the latter are idiographic and *verstehende*. Insofar as the cultural and the natural sciences can be shown to arise from the life-world, as Husserl's *Crisis* showed, and insofar as causal relations are already "essential categorical structures in the lifeworld" (76), these relations, which the natural sciences can clarify, are "legitimate categories for the human sciences" (see 73–75). These causal relations are experienced particularly at the level of elementary understanding basic to practical life, which lies at the base of all lifeworlds and which must come to terms with the natural environment, in particular that environment's causal processes that appear and are at work in: raw materials, tools, new technologies of every sort, artifacts, human consociates, regular causal sequences of events, the seasons, the movements of celestial bodies, the disturbing and destructive powers that can interrupt practical action, and organic processes (including those of one's own body). Elementary understanding provides the ground floor, as it were, to which contemplative first-order higher understanding (e.g., lifeworld cults and myths) turns its attention; furthermore, higher-order secondary understanding (e.g., in cultural sciences, including history), which takes account of such first-order higher understandings, cannot afford to neglect elementary understanding and the causal processes elementary understanding encounters in brute nature. Furthermore, to understand "what really happened" (vi, 95, 265)[1]—which is the target of historical science—in previous cultures' experiences of brute nature, it behooves history to consult with the natural sciences and not merely first-order higher understanding accounts, which, in pre-scientific contexts, might assign the origins of natural events to the will and power of spirits, demons, or the gods (See, e.g., 5–6, 51–52, 56, 73–75, 77–78, 95, 262, 264, 266, 268–270, 274, 314, 390, 392–393). That history, a matter of secondary higher understanding, rests atop layers of understanding that in the end suppose a lifeworldly pragmatic encounter with the natural environment, which implies for Seebohm that one cannot dispense with a grasp of the causal processes at work, and to construe history as a *verstehende* science in opposition to an *erklärende* natural science is to misconceive its nature (see, e.g., 60–64, 73–79, 268–270, 314, 390).

[1] Seebohm repeatedly defines history as an effort to determine "what really happened."

History needs the contributions of the natural sciences for additional purposes besides the clarification of the encounter with brute nature at the level of elementary understanding. Historians after the Enlightenment, which looked upon past eras as needing to be disabused of their irrationalities, saw the need for philology and archaeology, which attained scientific status in the nineteenth century. These sciences were seen as necessary to understand past cultures on their own terms since these sciences were able to check whether grammatical or linguistic corruptions had been introduced into past texts and to determine via chemical analyses (and paleographical methods) the age and material of texts, art, architecture, monuments, and artefacts. Many of the archaeological methods, of course, would be particularly useful in studying pre-historical cultures, namely, those without literary traditions. Causal explanations, such as those provided by the natural sciences, can also be used to give an account of changes in the natural environment of past cultures, including those that may have been humanly produced, as well as of alterations in the human body (e.g., epidemiological changes, malnutrition, and diseases) or in techniques and technology (e.g., windmills) to deal with nature—all of which could have had significant impacts on society at large, economics, and political life and all of which could help in the production of a critical historical science that would not be limited to historiography, which simply relies on the narratives handed down from past groups. Finally, historical reconstructions require the temporal framework of the calendar determined by the movement of celestial bodies in a Galilean framework and must presuppose as a background natural history, that is, the history of the universe, the geological history of Earth, and the history of evolution—all of which depend on the methods of the natural sciences (see, e.g., 68–69, 105–106, 110, 257–260, 261–265, 402).

But the convergence of history with the natural sciences is not simply a matter of layers of understanding and their presuppositions or ways in which the natural sciences can assist historical science; the method of prediction, which the bifurcated view of the natural and cultural sciences assigned only to the natural sciences, can also be used in the systematic cultural sciences, according to Seebohm. He argues, first of all, that analogues of the method of the natural sciences (e.g., methods of experimental research and mathematic models of predictions and hypotheses) can be applied in the systematic cultural sciences, though he also admits, in agreement with Felix Kaufmann (who denied any ontological differences between the objects of the cultural sciences and natural sciences), that such predictions will lack a degree of precision because of their complexity and the frequent emergence of new factors in the systems of social interactions. However, Seebohm also recognizes that participants in social interactions in elementary understanding make use of in-order-to motives (or purposes or final causes), which guide their own and their co-participants' use of efficient causal means, in order to understand their partner's actions and to develop expectations and predictions (which can be disappointed) about what these partners will do. For social scientists to attain a degree of predictability, they need to produce a methodologically-guided secondary understanding of these elementary understanding in-order-to motives and to test the predictions implied in these social scientific interpretations of elementary interpretations. Seebohm believes that such predictions can be enhanced when they deal with rational behavior that would be directed by an understanding of causal relations in social interactions. Such predictions are possible in the social sciences even though Seebohm, relying on Husserl's analysis of the abstractions for the natural sciences in the *Crisis*, in particular the first abstraction, which opens the domain of the life sciences and which admits only efficient and formal causality, recognizes that these sciences would rule out the final causes that are of "essential significance for causal explanations and predictions in the systematic human sciences in general, and especially in the social sciences" (313). In conclusion, prediction is possible in the systematic cultural sciences, despite the insistence by the bifurcators that the natural and cultural sciences are *toto coelo* different because the latter are not supposed to venture predictions, but it is a kind of prediction that differs from that found in the natural sciences, including the life sciences. Of course, predictions are relevant to the systematic cultural sciences but not to historical science, which preoccupies itself with the past, but Seebohm's argument about prediction should be seen as allied with the argument, on the basis of the practice of historical science, that the natural-cultural science bifurcation is problematic (see, e.g., Seebohm 2015, 7, 210, 217, 222, 230, 241, 279–280, 287–289, 312–313).

In Seebohm's view, Husserlian phenomenology has promoted, however, the gap between the natural and cultural sciences by arguing that the cultural sciences have an ontological priority over the natural sciences. While Husserl's *Crisis* posed the paradox of subjectivity in that the transcendental subject is both that with reference to which all other being exists and yet also itself an object among other objects in the world, Seebohm tries to eliminate the paradox by refusing to pursue the discussion on ontological grounds and instead by proposing the epistemological view that the subject and its objects (including its body) are given differently, in oblique intentions and direct intentions, respectively, and objects, independent of the subject, are also given for an intersubjective community to which subjective consciousness also belongs. The later Husserl, however, construed the ontological region of the natural sciences as a de-souled world in contrast to the ontological region of the human sciences, which was a spiritual world, from whose purposes and values the natural scientist must abstract. Husserl's differentiation between the realm of the natural sciences that bans "purposes" and the cultural sciences that belong to the realm of spirit, which takes ontological precedence over the material world, makes it questionable how the cultural sciences can now be empirical sciences. Furthermore, since Husserl's division requires the cultural sciences to validate their interpretations through interactions and communications, it now becomes questionable how cultural scientists can be disinterested, since they would have to enter into interactions that might jeopardize their objectivity, as we shall see below (see, e.g., 39–43, 92–94).

Schutz and Dilthey defended the viewpoint that the cultural scientist is only interested in second-order understanding of elementary and higher-order understandings of others and is willing to exclude all other aspects of the lifeworld (e.g., the engagement with brute nature). Seebohm considers their strategy as an effort to demarcate the province of the cultural sciences and to begin to specify what is involved in being a disinterested cultural scientific observer. However, Seebohm himself proposes, again, not to accept an ontic interpretation of the phenomenological reduction but an epistemological one that will make it possible to explore precisely how historical observers cooperate with the natural sciences and also achieve disinterestedness by means of the methodologies of history that are supported by the distance between historians and the cultures they study. Finally, Seebohm interprets "historism," that is, the taking the spiritual world as being ontologically prior to the natural world, as simply inverting "naturalism," which attempts to reduce all phenomena belonging to the cultural sciences to physics, though Seebohm rejects as absurd the idea of explaining historical facts or literature through brain physiology. Historism and naturalism simply play out the paradox of subjectivity, with the subject being, on the one hand, the source of all that is given in experience (historicism) and, on the other, being itself only one object among other objects (naturalism). Seebohm concurs, however, with Husserl's earlier opposition to *historicism* in which he argued on epistemic grounds that to argue for the historical or natural causation of all thought ends up undermining any claims that one might make, as would be the case if the history of science would end up undermining the view that the history of science demonstrates that the natural sciences can arrive at objectively valid knowledge. Seebohm contends further that to justify the superiority of the spiritual world over the natural world, Husserl, with Eugen Fink, resorted to a Fichtean-type, speculative, metaphysical, ultimate grounding in an absolute ego, beyond transcendental phenomenology as a philosophy of reflection. Seebohm, however, believes that such metaphysical speculations are irrelevant for the epistemological phenomenological reflections on the sciences and the system of the sciences that he presents (see, e.g., 94, 379–390).

One final result of the division between the natural and the cultural sciences is that one might construe social facts as purely spiritualized—a position that could be attributed to Schutz insofar as he, according to Seebohm, affirms that the human sciences should be reduced to second-order understanding, excluding all other aspects of the lifeworld. By contrast, Seebohm seems favorably disposed to Kaufmann's view that posits no ontological differences underlying diverging methodologies and asserts that social facts are psychophysical, though he suggests the appropriateness of Schutz's strategy of not pursing a transcendental division of the sciences but of instead reverting to the lifeworld to provide a foundation for all possible, higher-order, cogitative types (see, e.g., 94, 279–281, 292).

3.2 A Response to Seebohm's Overcoming of the Natural Science/Cultural Science Divide

Seebohm attributes two purposes to history as a cultural science: it seeks to ascertain what actually happened and to provide second-order understanding of the first-order understanding of others in past cultures (See, e.g., 262, 393; vi, 95, 265). Indeed, these two purposes would appear to be closely interrelated, since it would seem to be impossible to understand the entirety of what actually happened without understanding how those who experienced occurring events interpreted them. Furthermore, to suggest that what actually happened in some cases was that first-order understandings misunderstood events occurring in nature would presuppose that one knows what was actually going on in nature, which the natural sciences could clarify, as well as that one has a second-order understanding of these first-order understandings, which were in fact misunderstandings—a fact that itself can only be identified by comparing first-order understanding with what happened in nature. Seebohm's view of layers of understanding in a pyramid, rooted at its base in the encounter of lifeworldly elementary understanding and brute nature makes sense, and it rightfully suggests that for history to achieve its purposes, it must make use of the causal explanations provided by the natural sciences. Furthermore, it seems that it would be impossible to obtain an accurate second-order understanding of the first-order understandings if it were not possible to authenticate texts and to date the materials supposedly pertaining to a past culture and to explain that culture's natural environment including epidemiological facts, natural scientific accounts of the extant materials at hand and living organisms, and more remote geological and evolutionary factors at play. One would have to say that Seebohm is correct in conceiving History (with a capital H) as designating a complex multidisciplinary project that would encompass and require interpretative as well as causal-explanatory methodologies and findings in order to achieve its own purposes—in contrast to the neat divisions that Rickert and Windelband, threatened by natural scientific reductionism, hoped to uphold.

Nevertheless, Seebohm's multidisciplinary canopy seems to imply that the sciences under it retain their distinctive methodologies: archeology, paleography, and interpretive history (lower case) do not cease to apply their own distinctive methodologies. Hence, Seebohm recognizes that under the canopy of (multidisciplinary, upper case) History, there are limits to what natural scientific methodologies can achieve, for instance: that one cannot replace the methodology of philological-historical research with the methodology for natural scientific research (e.g., neurophysiology cannot help translate the *Iliad*), that different methodological procedures and foci characterize the history of nature and history as a cultural science, and that the predictions in the human sciences are more precarious than those in the natural sciences in part because the "objects" of the cultural sciences are actually "subjects" (See, e.g., 73, 259–265, 293–296, 302, 313, 380–381, 397, 408–409).

Not only do the natural sciences under the rubric of History retain their own methodologies, but, in fact, it can be argued that the many contributions the natural sciences can make to History serve the purpose of making more adequate, cultural-scientific interpretations of historical actors in past cultures. Dating monuments and manuscripts, authenticating texts via careful paleontology, explaining the brute natural events that past agents in their elementary understanding faced—all help to ensure the accuracy of one's interpretations of those agents. *It is as if the natural scientific methodologies are subordinated in History to the task of interpretation*. This becomes particularly clear in Seebohm's account of the development of history itself. Whereas the Enlightenment criticized and even rejected past cultural traditions from the perspective of the present through a kind of a-historical critique, those who insisted on understanding past cultures "in their own context," "immediate to God" as all cultures are, were also those who were the leading figures in the new sciences, philology, archaeology, and grammatical hermeneutics—before the later positivist attempts at natural scientific reductionism that Windelband and Rickert resisted by bifurcating the natural and cultural sciences. The very sciences that Seebohm sees as essential for the multidisciplinary History arose precisely to make possible a better understanding and interpretation of past cultural traditions than the Enlightenment permitted.

In the discussion of Husserl's prioritizing of the cultural sciences over the natural sciences, it is somewhat surprising that Seebohm should align Schutz with Dilthey in holding that:

> the human sciences are only interested in and reduced to the region of *secondary understanding* of first-order elementary and higher understanding of *Others*, i.e., the understanding of the meaning of immediate and fixed life expression of contemporaries and predecessors as observable objects. All other aspects of the lifeworld given in first-order elementary and higher understanding of the interpreter are excluded. (94)

In fact, Schutz shows himself quite open when it comes to the cultural sciences relying on other methodologies than merely a secondary understanding of other's meanings. For instance, Schutz recognizes that modern economics functions, not only with personal ideal types, but also with curves, mathematic functions, with the movement of prices, bank systems and currency, and that the statistical studies it makes use of have performed the "great" work of collecting information about group behaviors (1964, 84; see also 85). Schutz simply wants to hold out for the possibility the social scientist might shift the level of research to the individual actor when the problem at hand requires it. Furthermore, when Schutz explains how the cultural sciences construct ideal types, he argues for a postulate of rationality, which would include knowledge of the "different chains of means which technically or even ontologically are suitable" (80) for the accomplishment of an actor's ends and the knowledge of how these means might interfere "with other ends or other chains of means including their secondary effects and incidental consequences" (80). Seebohm rightly points out that Schutz neglects the difference between historiography and historical research; that he, with Husserl, neglected the methodology of philological-historical research developed in the nineteenth century by Schleiermacher and Boeckh; and that Husserl and Schutz did not appreciate sufficiently the contributions that the natural sciences might make to History. However, it seems impossible to me that, given Schutz's recognition of the role non-interpretative methods might play in the cultural sciences and of the importance of technology and the need for careful consideration of means-end rationality in the cultural sciences, he would have opposed the kind of assistance to the cultural sciences that the natural sciences could offer. In addition, he would not have excluded as irrelevant other aspects of the lifeworld given in first-order elementary and higher understanding (Seebohm 2015, 311, 320; see also Schutz 1964, 86–87; 1962, 116–117, 130–131; 1997, 144–145).

It is important to recognize that Schutz, and Dilthey, Rickert, and Kant, for that matter, were engaged in a struggle against positivism in the cultural sciences, as Schutz's engagement with Earnest Nagel and Carl Hempel in "Concept and Theory Formation in the Social Sciences" indicates (Schutz 1962, 48). Philosophers of the cultural sciences at the end of the nineteenth until the mid-twentieth century were engaged in a life and death struggle with positivism and may not have had the freedom to envision rapprochement with the natural science that the decline in positivism and the recent rise of the multidisciplinary "ethos" have made possible in our time. It is important to understand these predecessors within their own context. Given that context, it is remarkable to find in Schutz, despite his times, an irenic approach to the enrichment that mathematical models and technical means-ends analyses could offer to the cultural sciences.

It should also be said that Schutz converges with Seebohm's views on the matter of prediction in the cultural sciences, finding fault with the dichotomous view of the relationship between the natural sciences and the cultural sciences insofar as some cultural scientists might have resisted employing measurement, experiment, or prediction in the cultural sciences. Schutz espoused the idea that rational behavior of a constructed type in the cultural sciences would be by definition predictable, within the limits of what is typified in the construct. Such models of rational action could also be used to identify deviant behaviors. If the cultural sciences are based on constructs of the constructs of actors on the social scene, then Schutz's work could go a long way toward predictive possibilities since he placed more of a premium than most philosophers on the socially-learned and inherited typifications, including regularized patterns and recipes for action, which constitute the social world out of which cultural scientific constructs are formed and which enable us every day to operate in predictable manners, as can be seen, for instance, in the fact that "there is more than fair chance that a duly stamped envelope put in a New York mailbox will reach the addressee in Chicago" (1962, 56). In his essay on how typifications in everyday life make predictable the everyday behaviors that cultural scientific types capture, "Tiresias, or Our Knowledge of Future Events," Schutz reiterates his confidence in scientific prediction, while also highlighting the limits of predictability insofar as typifications are always general and the events they anticipate highly individualized, insofar as our stock of knowledge and prevailing interests are constantly changing, and insofar as unforeseen occurrences lying in the empty horizons of one's projection can induce alterations in the results anticipated. Even the because-motives from our past that influence our present choices for the future cannot deliver flawless predictability, since the determination of those motives is established by one's looking back in retrospect and choosing in accord with the relevances prevailing at the moment of looking back and liable to change at different periods of our lives. Like Seebohm, then, Schutz defends a place for predictability in the cultural sciences, one that is not as reliable perhaps as that to be found in the natural sciences, but one that undoes, to a degree, the split between the natural and cultural sciences (1962, 45, 48–49, 56–59; see also, Schutz 1962, 45–46; 1964, 282, 286, 287–288, 290; 1967, 95; 2013, 67–72).

Ontological questions play a role in the relationship between the cultural and the natural sciences. For Seebohm, Felix Kaufmann's opposition to ontological differences underlying the methodologies in the cultural and natural sciences was aimed at preserving the psychophysical character of social facts whose denial is a fallout from the division between the natural and human sciences that Dilthey, Husserl, and other phenomenologists insisted upon. Whereas Husserl went on to prioritize the cultural sciences over the natural sciences, Seebohm speculates that Schutz's turn to the lifeworld (which for Husserl, too, was a psychophysical region) might afford an alternative to this dichotomous approach that separates the psychic and the physical. Seebohm is correct on this strategy of Schutz's, since, in reaction to the dualism between meaningful action and physical behavior that led the behaviorists to translate all talk of meaningful action into physical behaviors, Schutz insists that actors in the social world, such as a jury in a trial, understand the actions of others (e.g., whether the accused acted out of pre-meditated intent) through commonly understood typifications, without trying to penetrate into some introspective private sphere and with a degree of public controllability. Such constructs, which do not separate psyche and body, of course, are what the social scientists construct (see, e.g., Seebohm 2015, 279–281, 291; Schutz 1962, 56).

It is interesting in this regard that Schutz nevertheless does seem to presuppose ontological differences between the objects of the natural sciences and the objects of the cultural sciences insofar as he affirms that there is an "essential difference in the structure of the thought objects or mental constructs formed by social sciences and those formed by the natural sciences" (1962, 58). The world of nature does not mean anything to electrons and molecules, but in social reality, the cultural scientist deals with a world that is being interpreted by actors on the social scene, and hence the cultural scientist must construct constructs of the constructs of those actors—which does not preclude the attention to causal factors, predictability, and the contributions natural science can provide, for which Seebohm calls. Indeed, the case can be made that, insofar as Seebohm recognizes such items as the indispensability of interpretation in history and cultural sciences whose objects cannot be reductively analyzed by natural scientific methodologies, as the differences between natural history and history as a cultural science, and as the problems of prediction if those who are predicted to act a certain way know about being predicted to do so, he too at least implicitly recognizes these ontological differences.

Seebohm's concern about ontological differences seems to result from his opposition to the paradox of subjectivity that, he claims, Husserl sought to resolve ontologically and that led him with Eugen Fink to arrive at a quasi-metaphysical idealism to ground transcendental phenomenology. Seebohm proposes to resolve the paradox by developing an epistemic interpretation of the ways in which subjective consciousness and objects are given (in oblique intention or direct intention), and this approach avoids the paradox that results from ontologizing the differences. To be sure, Husserl's procedure in *Ideas II*, which separates into two different ontological regions material nature and spiritual nature, both of which are constituted with reference to spiritual nature, may end up supporting the dualism of psyche and matter that Schutz avoids because he takes for his ontological region the lifeworld in which no such bifurcation is experienced or called for. Without implying the psychic/physical dualism and without precluding the contributions the natural sciences can make to history and the cultural sciences, Schutz's account, though, still encounters a stubborn difference between "objects" that do not interpret their world and subjects that do—and Seebohm, too, encounters this same difference in several of the theoretical maneuvers he makes within his book. At a higher philosophical level and in the name of philosophical self-responsibility, one would eventually have to engage in a full-blown constitution of these different "objects" and consider their relationship to transcendental subjectivity and intersubjectivity. Likewise, attention needs to be paid to the spiritual nature, in relation to which material objects receive their meaning and through which all constitution is undertaken, as Husserl attempts to do at the end of *Ideas II*. At this level, one might face the paradox of subjectivity and have to inquire further whether a quasi-Fichtean metaphysics beyond the level of transcendental phenomenology is called for to ground it.[2] However, such discussions take place on a plane above what is necessary to provide phenomenological grounding for the cultural sciences, as Seebohm states (See, e.g., Seebohm 2015, 39–40, 86–87, 292, 384, 388, 390; Schutz 1967, 44; Husserl 1989, 311–316).

However, just because Schutz thought that the constitution of intersubjectivity within the transcendental sphere was not possible, it is not the case, as Seebohm suggests, that he rejected the transcendental phenomenological reduction (he used it throughout chapter 2 of the *Phenomenology of the Social World*). Furthermore, he believed that transcendental phenomenology could be useful for the development of transcendental philosophy (as Maurice Natanson suggests), and he was already involved inchoately in developing such a philosophy, as his reflections on temporality and destiny in Goethe's *Wanderjahre* indicate (see, e.g., Schutz 1966, 82; Natanson 1968, 120–121, 142; Schutz 2013, 335–336, 398–403; Schutz, 1967, 43–44).[3]

3.3 Seebohm on Disinterested Observers

As a prelude to the question of the possibility of disinterested observation, it is worthwhile to consider Seebohm's criticism of Schutz's approach to history, in particular his failure to distinguish sufficiently between the systematic cultural sciences and historical research. Schutz's phenomenological descriptions include descriptions of predecessors as a structural aspect of the social lifeworld, but, for Seebohm,

[2] As I read Husserl in his *Crisis*, he speaks of a second *epoché* beyond the first *epoché* in which the transcendental ego constitutes the world, and in the second *epoché*, the absolute ego is the one that is "the ultimate unique center of function in all constitution" and so constitutes even the transcendental ego. See Husserl (1970, 186).

[3] That Schutz conceived his own discussions in Chapter 2 of *The Phenomenology of the Social World* were conducted under the constraints of phenomenological reduction, see Alfred Schutz (2004, 129–130).

he does not treat adequately the historical world, and he sees history as an extension of cultural sciences. In Seebohm's view, Schutz is basically interested in the present and in the past insofar as it is relevant to the present. In particular, Schutz's attention, for instance, to the history of law has to do with how that past can be applied in the presently valid legal system (Seebohm 2015, 79–80, 82–83).

Seebohm insists on the distinctiveness of history as a science, whose methodology has not been sufficiently recognized by phenomenology or by Husserl and Schutz, and Seebohm resists assimilating history to the present cultural sciences. Developments in the nineteenth century realized that interpretation of past history requires that the past be understood on its own terms, independently of any applications in the present, including judgments about whether past texts are true or not. As opposed to earlier historians who mined the texts of antiquity to find an ideal of humanity applicable in the present of the interpreters and that rejected past texts as not true where necessary, historians in the nineteenth century incorporated the assistance of archaeology and philology, newly acknowledged as sciences, and pursued the path marked out by Schleiermacher's first canon of hermeneutics, namely, that texts needed to be understood not out of the context of the interpreter but out of their own context and their contemporary addressees. Historical research involves careful, critical study of fixed life expressions, texts, and monuments of past eras, as opposed to mere uncritically-accepted historiographies—a distinction that Schutz and Husserl neglected. In other words, historical research requires the separation of interpretation and application, just the opposite of the direction in which Schutz proceeds insofar as history is but an extension of the present cultural sciences, such as jurisprudence and law, in which historical findings can be applied (see, e.g., vii, 66–67, 68–69, 82–83, 94, 104–105, 107, 112, 311, 320).

Furthermore, Seebohm emphasizes the *temporal distance* that separates the interpreting reconstructions from the reconstructed reality; that historical research into past eras, disconnected with the present, clearly takes account of; and that enables one to see clearly that present historians of past eras do not share the suppositions of the past they study, in contrast to contemporary historians studying a relative recent past that overlaps with their own time. While the cultural sciences are understanding sciences, for Seebohm, historical human sciences seek to understand past cultures on the terms of their own era on the basis of a distinctive structural foundation: the temporal distance between the present and the past. The awareness of this distance is the "justification" (128) for the first canon to understand the past on its own terms, the "condition of the possibility" (128) of the implementation of that canon, and the "condition of the possibility" (128) for historians to be disinterested observers of the past. The distance "admits and requires" (156) the implementation of the first canon in historical research. Seebohm repeatedly asserts that as the temporal distance between the interpreters and the interpreted past shrinks, so also the ability to implement the first canon decreases and the status of the historian as a disinterested observer diminishes. When twentieth century analytic philosophy borrowed psychological generalized conditionals from everyday psychological experience for historical explanations, they overlooked how such explanations are hopelessly "contaminated" (144) by present economic, legal, or political conditions;

they neglected in effect the temporal distance and the first canon which depends on that distance (see, e.g., 95, 152, 156, 405). In Seebohm's opinion, Schutz's readiness to assimilate history with the systematic cultural sciences leads him to neglect the distinctiveness of history and the advantages afforded by the temporal distance and the practices historical science has developed for the prospects of disinterested observation in history. In addition, while Seebohm's discussion highlights the inbuilt protections for disinterested observation in historical science, it also poses serious questions about how those in the systematic cultural sciences will achieve disinterestedness and objectivity.

> Prima facie there are no methodological and epistemological problems because secondary understanding of immediate linguistic life expressions of Others in present communication "works" without the need to apply methodologically guided interpretations. The problem is, however, that interpretations in the historical human sciences are able to perform a methodological abstraction that is constitutive for the possibility of separating interpretation and application and thus of the cognitive attitude of "disinterested" interpretations in the historical human sciences. This distance vanishes gradually in contemporary history and completely in the secondary understanding of life expressions of Others in present communications. The question is, hence, how methodologically guided eliminations of misunderstanding and non-understanding in interpretations are possible in the systematic human sciences. (405, see also 311)

While positivistic and analytic methodologies favored the use of experimental methods in the cultural sciences that could be tested empirically, e.g., through predictions, which have their limitations, the proponents of the independence of the cultural sciences conceived the cultural sciences as a matter of secondary understanding of others, which requires alternative methods of validation. Unfortunately, according to Seebohm, Schutz said nothing about how the objective validity of interpretations was to be established. As regards disinterested observation and validation in the systematic cultural sciences, Seebohm argues that the systematic cultural sciences lack both the temporal distance and the first hermeneutic canon that ensure disinterested observation in the *reproductive* interpretations of historical science, but they have an advantage in having immediate access to observing the present life expressions of their "objects." When anticipations about how the observed party should act fail to be realized, cultural scientists can take such failures to indicate that they have not understood the other and to compel them to check their interpretations. In order to carry out such checking, the systematic cultural scientists need to develop *productive* (as opposed to reproductive, past-oriented) interpretations, which can be corrected for not-understanding and misunderstanding through communications in the present. However, Seebohm questions whether it is possible at that point, in such communications, for the observer/interpreter to be disinterested. Just insisting that the observer be disinterested is "not very helpful" (289) and attempting to ensure disinterest through the subjective attitude that the researcher ought to adopt seems "dubious" (293). Nevertheless, the observer continues to be "interested" in the objective validity of the interpretation of the action or the life expressions of the actor (279–281, 284, 286, 287, 290–291).

In line with such scientific efforts to be self-corrective, Schutz's postulate of adequacy requires that the social scientific constructs ought to be reasonable and understandable for actors and their contemporaries. At one point Seebohm categorizes the postulate of adequacy as a version of the requirement for the testability and the confirmation or disconfirmation of hypotheses and theories, in other words, for the objective validation of interpretations, which are confirmed if actors agree with them or disconfirmed if they disagree (319, 321). To determine such adequacy, though, one would have to make use of diagnostic questionnaires or interviews, which concern the in-order-to and because motives of actors and which are to be constructed independently of the interests of researchers. However, Seebohm believes that such processes can easily fall into dialogues about why it is good or reasonable to realize such purposes or disputes about what has been said or understood, and such dialogues introject a dimension of application (judgments about what is true [104]) into what is supposed to be a matter of merely interpreting another. Such an introjection leads to the results that the scientists may not understand the other on the other's terms and that the requisite observational disinterest is not preserved. In the end, Seebohm argues that the observer in the present cultural sciences cannot occupy a standpoint outside the structural whole of social relations in a concrete social lifeworld, as is the case for historian reconstructing a past social lifeworld. Consequently, scientific observations and interpretations will be determined by the context of the structures of pre-scientific social life and the theories current in the cultural sciences. As a result, "the inquisitive dialogue with the 'objects' will, therefore, necessarily include 'suggestive' questions" (295). Finally, since the "objects" of the systematic cultural sciences are subjects, interpretations of them can alter their behavior in the light of descriptions and predictions, thereby falsifying the scientific findings the minute they are conveyed to the "objects" studied (289–291, 293–294, 296, 298, 305–306).

To appreciate Seebohm's pessimism about disinterestedness in the systematic cultural sciences in contrast to his optimism about the historical sciences, a little more needs to be said about the dialogue between researcher and "object" that aims at establishing this validity, or adequacy, in Schutz's terms, of interpretations. First of all, Seebohm believes that possibility of such "diagnostic" dialogues is a basic assumption of the postulate of adequacy. Such dialogues, in contrast with more loosely-construed productive dialogues, are supposed to be the "warrants of the distance that is required for disinterested and objectively valid observations and interpretation" (408). Furthermore, the attitude of an interrogator in a pure diagnostic dialogue is the "presupposition of the possibility of separating interpretation and application in dialogues and communication in general" (405), and, in the end, if participants agree with a hypothetical interpretation of the cultural scientists, then that interpretation is confirmed, and, if not, it is disconfirmed. For Seebohm, the problem of developing methodologies for diagnostic dialogues is *the* basic problem of an epistemology of the social sciences. Dialogues in the cultural sciences differ from the diagnostic dialogues in psychology interested in participants' reaction to their social environment or in the elicitation of the traces of what is suppressed in psychoanalytic research, but, even here, Seebohm believes that misunderstandings leading to disputes are possible and that present psychological theory itself determines researchers in a way that might impede disinterestedness. Seebohm expresses different views of what might be involved in diagnostic dialogues in the systematic cultural sciences. In such interviews, cultural scientists are supposed to "explain their theories in 'simple' terms before asking whether their clients agree or disagree" (325). Such interviews can preserve disinterestedness only if questioning of participants is "restricted to inquisitive and diagnostic questions first about their in-order-to motives and then about their because motives" (290). In social psychology, the interrogator's knowledge of the types of structures of the social context needs to be bracketed to keep "the required distance of disinterested interpretations" (406–407). Finally, the "objects" of such interpretation can develop an interest in the interpretations or ideal types of their activities developed by their interrogators, and then the diagnostic dialogue is at an end and a *productive dialogue* will ensue with the participants drawing researchers into a common "search of truth" (408) or advancing the participants' own special interests on the basis of the research. Productive dialogues are required to eliminate misunderstanding and not-understanding (on the part either of the social scientist or the object interpreted regarding the scientist's questions, for example), but such dialogues impede the very distance required between researchers and their objects in the systematic cultural sciences since "there is no such distance in productive dialogues" (411; see also See also 290, 298, 298n34, 303–304, 305–306, 321, 324, 325, 324, 408).

3.4 A Response to Seebohm on Disinterested Observers

Seebohm is correct to point out that Schutz does not show awareness of the discussions in the early nineteenth century by Schleiermacher and Boeckh about such things as canons of hermeneutics, and consequently Schutz does not discuss directly the problem of the unity of interpretation and application, neither with respect to philological historical research or the science of law. Though Seebohm does mention that the issue emerges in the discussion of law for Schutz, he cites no sources in Schutz to substantiate this claim. In addition, as a result of his lack of familiarity with nineteenth century discussions, Schutz does not distinguish sufficiently between historiography and historical research and between the methodologies of the systematic cultural sciences and those of historical research (82–83).

Nevertheless, it would be inaccurate to suggest that Schutz does not recognize the significant differences between the past and the present, that he assimilates the past to the present, or that he would not be sympathetic with differentiating interpretation from application, in line with the first canon of hermeneutics that demands that we interpret the past on its own terms. Schutz's discussion of the structure of the social world, which Seebohm references but does not examine carefully, illustrates this point:

> The schemes we use to interpret the world of our predecessors are necessarily different from the ones they used to interpret the world. If I wish to interpret the behavior of a contemporary, I can proceed with confidence on the assumption that his experiences will be pretty much like my own. But when it comes to understanding a predecessor, my chances of falling short of the mark are greatly increased... I have no way of making sure that my own interpretive scheme coincides with my predecessor's expressive scheme when he made use of the signs in question. Satisfactory interpretation of signs used in the past is therefore always problematic. Think, for instance, how much controversy there has been over the "correct" interpretation of the works of Bach in terms of the "objectively given" system of music notation. Even the history of philosophy is teeming with disagreements over the proper interpretation of terms used by philosophers in the past. This uncertainty is different in kind from the uncertainty we have about words and other signs used by our contemporaries, for we can always ask the latter what they mean and so settle the question once and for all. (Schutz 1967, 210–211)

While the temporal distance and the first canon of hermeneutics assist in preserving disinterestedness and objectivity in historical research and provide advantages the systematic cultural sciences lack, Seebohm recognizes that even though the structures of intersubjective temporality and elementary understanding are in place, the first canon still needs to be applied to reconstructions and interpretations of contexts. In fact, historical research as practiced gives evidence that more is needed than the distance and canon by themselves insofar as the canon itself serves as an ideal up to which historians themselves may not have lived and for which they have had to strive. For instance, a colleague of mine, Dr. Michal Rozbicki, has argued in his award winning book *Culture and Liberty in the Age of the American Revolution* that even distinguished historians of the American revolution have operated with a self-evident, unproblematic concept of equal liberty as a norm against which the revolutionaries fell short, instead of understanding how the culture of the revolutionaries assumed social inequality and examining instead the socio-cultural cultural conditions and the protagonists who gradually modified this understanding of liberty (see Rozbicki 2011, 2, 8, 10, 18). Even with temporal distance and the presumably widespread understanding of the canon, historical researchers are capable of not appreciating agents of past eras and their contemporary addressees on their own terms. Even an eminent historian like Dr. John Hope Franklin, who wrote extensively on race relations in United States history, struggled to ensure that the many humiliations he had suffered as an African-American historian not inspire a hatred or revenge-seeking that might "pollute" his scholarly work; and Franklin admitted that "the task of remaining calm and objective is indeed a formidable one" (Faust 2015, 97; see also, Seebohm 2015, 112, 129, 294).

The missing ingredient, which is required but which even when added does not ensure disinterestedness or objectivity, is that the historian must adopt a hierarchy of relevances whose overriding purpose is "not to master the world but to observe and possibly to understand it" (Schutz 1962, 245), and must replace the interests of his or her biographical situation with those of the scientific situation. John Hope Franklin precisely describes the struggle involved in letting one's scientific relevances govern one's historical research. Alfred Schutz discusses in several of his works the effort to prioritize such relevances as involving a shift in attitude comparable to the phenomenological *epoché* and as belonging to the finite province of scientific theory whose cognitive style Schutz delineates in great detail (see, e.g., Schutz 1962, 245–253; Schutz 1962, 63; 1997, 143–144). To my knowledge, Seebohm does not consider carefully this discussion by Schutz, and his dismissive statements that it is not helpful to add that the observer ought to be disinterested and that discussions of the attitude of individual researchers are of dubious worth underestimate the importance of at least striving to carry on one's research in accord with predominating scientific interests. Such governing relevances may help in the complex dialogue that researchers in the systematic cultural sciences and the subjects they investigate carry on and that are aimed at establishing the validity (which Seebohm equates with adequacy) of the scientific results (Seebohm 2015, 289, 293).

Although Seebohm provides various versions of how cultural scientists ought to proceed in such dialogues, it does not seem necessary that researchers explain their *theories* in simple terms to their subjects. It does seem, however, more appropriate that researchers deploy inquisitive and diagnostic questioning about the in-order-to motives of their subjects. Although Seebohm fears that such questioning will easily slip into a productive dialogue that might include questions about whether the purposes of the subjects are good or reasonable, that could consist in "suggestive" questions or the injection of scientific theories on the part of the researchers, or that might see both subjects and researchers succumbing to a disputative search for truth together, it is not necessarily the case that in interviewing subjects the cultural scientist is "on the same level" with the subject researched. This is because, even in the dialogue between them, the cultural scientist ought to be guided by the in-order-to motive of simply trying to find out what the action means to the actor, what the actor's in-order-to motive might be, and so even if the subject objects to the interpretation of the researcher—that objection becomes further evidence that the cultural scientist needs to consider dispassionately. The scientist may have misunderstood the subject, the subject may have misunderstood the scientist, or perhaps the subject is incapable of recognizing the motives at work in himself or herself. The scientist, because of the scientific relevances adopted, must consider with fairness and with an interest in arriving finally at the truth, whatever response the subject makes to proposed interpretations, and in this the scientist differs from the subject who does not adopt the relevance framework of science. Interpretations are not necessarily confirmed or disconfirmed depending on whether the subjects agree or disagree with them. Cultural scientists, though, ought not to endeavor to convince the subject that his or her purposes are or are not good or reasonable and ought not to be competitively striving to ensure that their scientific interpretations prevail over the subject's objections. All along, cultural scientists seek to determine what the actions mean from the point of view of the actor by adhering to the relevances that science mandates, which may not absolutely guarantee objectivity, but which at the least play a greater role in moving in that direction than Seebohm acknowledges (142).[4]

[4] Schutz explains how relevances are crucial for disinterestedness here.

Another difficulty is that Seebohm seems to equate the postulate of adequacy with the determination of whether one's interpretations, hypotheses, or theories are objective, or valid. In the "Ineditum" that Seebohm cites, when Schutz turns to the second postulate after the postulate of subjective interpretation, he notes regarding the subjective meaning of the actor that "its concept and theory have to be objective in the sense that they are subject to customary control by fellow scientists in several respects" (145). For this purpose, Schutz goes on in that same paragraph to introduce the postulates of logical consistency and adequacy. As I read Schutz, then, the postulate of adequacy, like the postulates of subjective interpretation and logical consistency, are intended as postulates that ought to shape the preparation of one's theories, hypotheses, and interpretations before they are presented to the scientific community whose final assessments will determine whether they are objective or valid. Lester Embree, concurring that adequacy does not produce verification, argues that it, in fact, ensures the "plausibility" of one's interpretations (2015, 11). In fact, repeatedly, throughout the "Ineditum," Schutz makes it clear that fellow scientists are the ones to determine whether the cultural scientist's findings are verified or not. Lester Embree rightly summarizes what the cultural scientist's constructs would agree with in order to be valid when he mentions that "social-scientific constructs ultimately refer as well to the subjective meaning understood by the actor in her subjective interpretation and that is what social-scientific constructs, if true, coincide or agree with" (2015, 169; see also, Embree 2015, 28, 133, 148, 149).[5]

By emphasizing that verification takes place in the effort to establish adequacy between researcher and subject instead of within the exchange between fellow scientists, Seebohm has misplaced the site of validation in Schutz. In doing this and in neglecting the importance that the scientist's attitude be directed by scientific relevances, Seebohm makes the entire weight of the process of finding objective truth rest upon the final dialogue between researcher and subject. And that dialogue itself depends essentially on how well the researchers can keep their everyday relevances, their theoretical suppositions, and their tendency to evaluate rather than describe under control, that is, out of the process—a task made much more difficult by the fact that cultural scientists do not have the temporal distance and canons of hermeneutics that make the separation of interpretation from application more easily achieved in historical sciences.

[5] It is important to note that Schutz, on p. 148, mentions that verification happens in the cultural sciences in the same way that it does in the natural sciences, as long as we realize that empirical observation is not merely sensual perception of occurrences of the outer world. Verification in the natural sciences occurs when the community of scientists finds a hypothesis valid. See Embree (2015, 169). Embree, in personal comments, stated that Seebohm was a falsificationist after the fashion of Karl Popper, whereas Seebohm thought that Schutz was a verificationist. Given that Schutz in "Tiresias, or Our Knowledge of Future Events" (1964, 286), interpreted Husserl's idealization of "and so forth and so on" as implying the assumption "valid until counter-evidence appears," one might be able to build an argument on the basis of his understanding of everyday life typifications, that Schutz was actually a falsificationist in the sciences also. We will abide by the use of "verificationist" in the text of this paper to indicate that one's research must pass the test of the community of scientists. "Passing the test," though, could simply mean that no sufficient counter-evidence has been advanced.

The problem here is that Seebohm seems to conceive as only detrimental to our interpretation of others what Heidegger calls our unthematized totality of involvements, or *Vorhaben*, and what Schutz and Husserl might describe as the system *of* sympathies, sensibilities, or immediately applied typifications working on a level of passive synthesis beneath the control of the rational "I." Though it is indeed possible that these subterranean suppositions can lead us to impose our own interpretations on others, it is also possible that they can assist us in understanding better those who differ from us. For instance, it is conceivable that an anthropologist with religious convictions of her own might be able because of those convictions to understand contemporary fervent Muslim groups better than secularist scientists, who, in turn, might better understand secularist groups who have struggled against religious domination in places like Turkey or Iran. Of course, the adoption of the scientific attitude, consultation with the subjects studied (as one among other postulates to be observed prior to the formulation of one's results), and the testing of the scientific community would still be needed as part of a comprehensive project of seeking objectivity in addition to our interpretive sensibilities, which can be allowed some free play if we have these other safeguards in place. Seebohm's approach to validation and disinterestedness seems to have an "ascetic" tone to it: one needs to remove or suppress any interests in oneself that might lead one to impose interpretations on others. This tone, characterizing Seebohm's understanding of objectivity, appears when he characterizes analytical philosophers who used generalized psychological conditionals of everyday experience for historical explanations as producing explanations that are "hopelessly contaminated" (Seebohm 2015, 144) by present conditions. The effort to provide valid interpretations in the cultural sciences can be described, not only negatively, as trying to prevent our cultural suppositions from sneaking into our work and imposing themselves on others, but also positively, as endeavoring to understand others as accurately as possible, or as Schutz defines the final, overriding relevance of theory: "Its aim is not to master the world but to observe and possibly understand it" (Schutz 1962, 245). At one point, in contrast to the usual ascetical tone of his approach to objectivity, Seebohm seems to move toward a more positive approach to objective interpretation when, in criticizing attitudinal approaches as a route to disinterest, he states, "Scientific observations are interested. Their interest in the human sciences is precisely the objective validity of the interpretation of the action or life expression" (2015, 289). In fact, if all validity does not rest in the end on a concluding interview between the subject and a cultural scientist who must be purely disinterested, but rather depends on comprehensive safeguards, relevances (including the positive relevance of striving to understand accurately another), the implementations of the postulates (as a prelude to presenting findings), and finally entrusting one's results to one's fellow scientists to establish ultimate validity, one would be able to release and give freer rein to those hidden sensibilities not under our control but nevertheless linking us to those we study (Heidegger 1962, 191).

Finally, Seebohm may be correct in noting that when cultural scientists acquaint their subjects with their results, predictions, and hypotheses, it is possible that subjects described, because they are conscious agents, are able to alter their behavior, falsifying the interpretations or predictions scientists have developed. This, of course, might provide the starting point for new investigations, showing how the relaying of scientific information alters everyday actors' behavior. Or perhaps, one simply needs to acknowledge that the findings one arrives at always bear a temporal index and can always be modified by the passage of time.

3.5 Schutz and Seebohm

There are several areas in which one working from within a Schutzian perspective might take exception to Seebohm's statements. For instance, Seebohm criticizes Schutz for conflating the lifeworld with the natural attitude, and while it might make sense to say that the mathematical entities of higher algebra are given in the natural attitude, that is, before the reduction, it does not make sense for Seebohm to say that they are given in the lifeworld, presumably because they are much too theoretical in nature for the practical, everyday lifeworld. However, for Schutz, the provinces of meaning—from the world of working to the theoretical province of meaning, the latter of which requires a peculiar *epoché*—all belong to the lifeworld; hence mathematical entities would be given in the theoretical province of meaning, which itself belongs to the lifeworld, but not to the pre-theoretical world of working (Seebohm 2015, 287 n. 22). In addition, when Seebohm discusses "because motives," he presents them as actors' "applied knowledge about the means for realizing the purposes guiding their in-order-to motives" (290). He is, in fact, interpreting them as the knowledge of how to engage causally with objects in the world in order to realize one's purposes. In fact, because motives result from speculating about the motives in the past that acted on one from behind, motivating one even to choose one's in-order-to projects, which are entirely future in their orientation. Once I have decided on a project or realized it, I stop and reflect about those events and persons, my because motives, which in the past before my now past (in pluperfect tense) decision or action may have influenced those past decisions or actions. Moreover, Seebohm's argument that eidetic methodology cannot operate independently of the phenomenological reduction deserves critical scrutiny insofar as Schutz's *The Phenomenology of the Social World* indeed engages in eidetic reflection within the natural attitude, appealing methodologically to Husserl's "Nachwort zu meinen 'Ideen'" (Schutz 1967, 43–44, 44 nn.85, 86, 88, 89; Seebohm 2015, 16, 26). Besides, Husserl himself did not seem to object to Schutz's separation of eidetic analyses from the reduction insofar as he had nothing but praise for Schutz's *The Phenomenology of the Social World* (Michael Barber 2004, 42).[6] Finally, Seebohm seems to criticize Husserl for not attending sufficiently to the interruptions of communication and to the mechanisms that might lead to successful communication. However, a case could be made that Husserl's account of empathy, which relies

[6]The text cites a letter to Schutz from Husserl dated May 3, 1932, and praising Schutz's *The Phenomenology of the Social World.*

extensively on the widespread deployment of passive syntheses, really develops the roots of communication and the origin of a common world (shared even with animals) against the background of which all communicative difficulties can be resolved (Seebohm 2015, 282; see also Husserl 1973a, b, c.).

Seebohm has clearly shown, though, how Schutz, as well as Husserl, neglected the important developments in historical research and hermeneutics in the nineteenth century. He points out ways in which history and the cultural sciences need to be open to contributions from the natural sciences, and I have suggested that Schutz, despite his opposition to positivism, showed himself even in the 1950s to be open to possible multidisciplinary collaborations with the natural sciences. Finally, I have suggested that Schutz's notion of disinterested observation was more comprehensive than Seebohm seemed to recognize, and that reliance on Schutz's notion might lead to a less ascetic and more positive notion of the cultural scientific project.

References

Barber, Michael. 2004. *The participating citizen: A biography of Alfred Schutz*. Albany: State University of New York Press.

Embree, Lester. 2015. *The Schutzian theory of the cultural sciences*. Dordrecht: Springer.

Faust, Drew Gilpin. 2015. John Hope Franklin: Race and the meaning of America. *The New York Review of Books*, December 17.

Heidegger, Martin. 1962. *Being and time*. Trans. John Macquarrie and Edward Robinson. San Francisco: Harper San Francisco.

Husserl, Edmund. 1970. *The crisis of the European sciences and transcendental phenomenology: An introduction to phenomenological philosophy*. Trans. David Carr. Evanston: Northwestern University Press.

———. 1973a. *Zur Phänomenologie der Intersubjektivität: Texte aus dem Nachlass, Erster Teil, 1905–1920*, Husserliana XIII, ed. Iso Kern. The Hague: Martinus Nijhoff.

———. 1973b. *Zur Phänomenologie der Intersubjektivität: Texte aus dem Nachlass, Zweiter Teil, 1921–1928*, Husserliana XIV, ed. Iso Kern. The Hague: Martinus Nijhoff.

———. 1973c. *Zur Phänomenologie der Intersubjektivität: Texte aus dem Nachlass, Dritter Teil, 1929–1935*, Husserliana XV, ed. Iso Kern. The Hague: Martinus Nijhoff.

———. 1989. *Ideas pertaining to a pure phenomenology and to a phenomenological philosophy, Book 2: Studies in the phenomenology of constitution*. Trans. Richard Rojcewicz and André Schuwer. Dordrecht: Kluwer Academic Publisher.

Natanson, Maurice. 1968. *Anonymity: A study in the philosophy of Alfred Schutz*. Bloomington: Indiana University Press.

Rozbicki, Michal Jan. 2011. *Culture and liberty in the age of the American revolution*. Charlottesville and London: University of Virginia Press.

Schutz, Alfred. 1962. Common sense and scientific interpretation of human action. In *Collected papers 1: The problem of social reality*, ed. Maurice Natanson. The Hague: Martinus Nijhoff.

———. 1964. *Collected papers 2: Studies in social theory*, ed. Arvid Brodersen. The Hague: Martinus Nijhoff.

———. 1966. *Collected papers 3: Studies in phenomenological philosophy*, ed. Ilse Schutz. The Hague: Martinus Nijhoff.

———. 1967. *The phenomenology of the social world*. Trans. George Walsh and Frederick Lehnert. Evanston: Northwestern University Press.

———. 1997. Positivistic philosophy and the actual approach of interpretative social science: An ineditum of Alfred Schutz from spring 1953, ed. Lester Embree, *Husserl Studies* 14: 144–145.
———. 2004. *Der sinnhafte Aufbau der sozialen Welt: Eine Einleitung in die verstehende Soziologie*, ed. M. Endress and J. Renn, vol. 2 of the *Alfred Schutz Werkausgabe*. Konstanz: UVK.
———. 2013. *Collected papers 6: Literary reality and relationships*, ed. and trans. Michael Barber. Dordrecht: Springer.
Seebohm, Thomas M. 2015. *History as a science and the system of the sciences: Phenomenological investigations*. New York: Springer.

Michael D. Barber Ph.D., Professor of Philosophy at Saint Louis University, received his Ph.D. from Yale University in 1985. He is the author of seven books in the area of the phenomenology of the social world, including *The Participating Citizen: A Biography of Alfred Schutz* which won the Ballard Prize in 2007. In addition, he has published over 80 articles in this area of phenomenology. He has published in *Husserl Studies* and *Human Studies*, and his papers appear in collections published by Routledge and Oxford University Presses. He has held leadership positions in various phenomenological organizations, including the editorship of *Schutzian Research*.

Chapter 4
Seebohm's Hermeneutics

Robert Dostal

Abstract This essay considers Seebohm's philological hermeneutics in relation to the philosophical hermeneutics of Heidegger and Gadamer. Though the hermeneutics of Seebohm and Gadamer appear to disagree, this essay points to five fundamental characteristics which they share. Their primary disagreement concerns the so-called first canon of hermeneutics which requires that we distinguish the context of the interpreter from the context of the text. A corollary of this canon is the distinction between the meaning and the application of a text. Though there are aspects with regard to this canon about which they agree, Gadamer dismisses this canon and Seebohm with important qualifications accepts it. Nonetheless, Seebohm argues that he is not in direct disagreement with Gadamer and his hermeneutics since Gadamer is concerned primarily with truth and Seebohm with the validity of the interpretation.

4.1 Introduction

Thomas Seebohm was engaged with hermeneutics for his entire academic career. In the 1970s he published his critique of hermeneutical reason and a lengthy study of the development of philosophical and scientific thought in early Russia (1972, 1977). More recently he published *Hermeneutics: Method and Methodology* (2004). And, finally, he left a completed manuscript on history and the philosophy of science in which philology and hermeneutics play a central role (2015). History according to Seebohm necessarily presupposes what he calls "philological-historical interpretation" (2015, 125). In short, history as a science presupposes hermeneutics.

In this paper I continue a private and public conversation with Seebohm that goes back to the 1970s when I studied with him at Penn State. The public conversation

R. Dostal (✉)
Bryn Mawr College, Bryn Mawr, PA, USA
e-mail: rdostal@brynmawr.edu

© Springer Nature Switzerland AG 2020

T. Nenon (ed.), *Thomas Seebohm on the Foundations of the Sciences*, Contributions to Phenomenology 105,
https://doi.org/10.1007/978-3-030-23661-8_4

came in response to the publication of his *Hermeneutics* book.[1] My talk and subsequent publication considered Seebohm's hermeneutics in relation to the hermeneutics of Gadamer. In his response to my paper, Thomas states explicitly that over time "I changed my understanding of *Truth and Method.*" He called my account "justified and precise" and then went on to say that I overstated his disagreements with Gadamer and the significance of Gadamer for his work. In fact, he had no quarrel with Gadamer.

Here I consider Seebohm's hermeneutics ("philological-historical hermeneutics" as he sometimes calls it) in relation to the philosophical hermeneutics of Heidegger and Gadamer. The contrast will help sharpen our understanding of what is specific to Seebohm's hermeneutics. On its face, it seems as though these two versions of hermeneutics are quite at odds. Seebohm tells us that they are not. How not? And why not? This leads us to consider most carefully Seebohm's defense and usage of what might be called the "first canon" of hermeneutical philology—that a text is to be understood in its own context, distinct from the context of the interpreter. A corollary to this canon is the separation of interpretation and application. This takes us to a consideration of Seebohm's concerns about ontology in Husserl and the question of an ultimate foundation, *eine Letztbegründung*.

4.2 Seebohm's Hermeneutics in Relation to "Philosophical Hermeneutics"

4.2.1 Their Agreements and Disagreements

There appears to be a large and fundamental disagreement between Seebohm's hermeneutics and Gadamer's hermeneutics. Seebohm consistently refers to Gadamer's hermeneutics as "philosophical hermeneutics" and refers to his own hermeneutics in various ways—sometimes "methodological hermeneutics" (2004, 55) or "philological hermeneutics." Of course, what both Gadamer and Seebohm are doing in their own ways is providing a phenomenology of interpretation. Gadamer is not particularly forthcoming about this because of his hesitancy to identify with Husserlian phenomenology, but in the "Foreword" to the second edition of *Truth and Method* he acknowledges that "my book is phenomenological in its method" (1999, xxxvi).

In spite of the fact that Gadamer here (clearly pressed by critics) refers to the "method" of his work, *Truth and Method* is rightly notorious for abjuring method. Seebohm embraces method. An aspect of Gadamer's taking a distance from method

[1] In the fall of 2005 Lester Embree organized a panel discussion of *Hermeneutics: Method and Methodology* at SPEP. The papers by Lester Embree, Thomas Nenon, and myself were subsequently published in the *International Journal of Philosophical Studies* 16 (December 2008), 719–769, with responses to each paper by Thomas Seebohm.

is his criticism of nineteenth century methodological hermeneutics as can be found in Schleiermacher, Boeckh, Ast, and others. In contrast, Seebohm builds on their work.

Seebohm turns to Husserlian phenomenology to provide an account of interpretation, though he, too, is critical of some aspects of Husserl's work including its lack of any serious consideration of the scientific status of history and philology (2015, 281). Accordingly, Seebohm's task in his reliance on Husserlian phenomenology is not to articulate for us what Husserl's hermeneutics was but rather to develop a hermeneutics on the basis of Husserlian phenomenology. For Seebohm, a phenomenology of hermeneutics, that is, a phenomenology of philology or "historical-philology" is both descriptive of the practice of philology and critical of its methods. It both justifies and clarifies. It is a "critique" of philology in the Kantian sense. In this regard Seebohm identifies the Kantian critical project with the Husserlian phenomenological project.

At the heart of this difference between Gadamer's phenomenological treatment of hermeneutics and Seebohm's are their respective treatments of the so-called first canon of methodological hermeneutics that requires that one distinguish between the context of the work interpreted and the context of the interpreter. As mentioned already, a corollary of this canon is the distinction of interpretation and application. One is to interpret the work in its own context and then, in a distinctly separate operation, consider how it might be applied in one's own context. Gadamer argues simply that to interpret is, at the same time, to apply. He denies the distinction.

But before we consider more carefully why and how Seebohm and Gadamer disagree about this, we should note a good bit of continuity in their respective accounts of interpretation. Here are five important features of the account of interpretation that they largely share, though in each case Seebohm provides some important qualifications with regard to his version of the feature—only some of which will be considered here. These qualifications do not represent a fundamental redefinition of these features, but they do provide a much more nuanced account of the role of these features in interpretive practice.

First, Seebohm and Gadamer both embrace the second canon of interpretation—its circularity. Any text can and should be understood in its context. The text is only a part within a larger whole. We understand the whole in terms of its parts and the parts in terms of the whole. It's an iterative and ongoing process that, in principle, never ends. The relationship of whole and part can be deepened and expanded. This means importantly that no interpretation can be considered final. Gadamer objects to apparent claims to scientific certainty and proof in nineteenth century methodological hermeneutics. Seebohm argues that the ineluctable circularity of interpretation meant also for the nineteenth-century methodologists and for himself the lack of finality in interpretation. He argues further that, strictly speaking, the second canon is not a methodological principle but rather a heuristic (2004, 162–167, 169–218). As such, it is more a matter of prudence and discipline than method. This treatment of the canon is close to that of Gadamer who espouses discipline rather than method.[2]

Second, both Seebohm and Gadamer deny that the intention of the author is a determinant of the correct interpretation. Both acknowledge that a text may say more than the author intended or understood. Nonetheless Seebohm argues that the intention of the author is relevant as a part of the context of the text, and he has some suggestions with regard to the difficulties of determining the intention of the author (2015, 152).

Third, the concept of the "eminent" text and its connection to the concept of genius is important for both accounts. These concepts are part of the nineteenth-century methodology for philology. Briefly put, an eminent text is a work of genius that violates the usual rules, most importantly the rules of genre. Seebohm acknowledges that there are such texts. They pose difficult problems for interpretation because genre is so important for interpretation. Seebohm criticizes Gadamer for "privileging" the eminent text as the model for understanding the hermeneutical situation. Seebohm reminds us that most texts are not eminent.

Fourth, for both Seebohm and Gadamer the paradigm case for hermeneutics is the written text. The text is in a language. Linguisticality is a basic feature for hermeneutical understanding for both. Seebohm is concerned that Gadamer's proposition that "*Sein, das verstanden werden kann, ist Sprache*" [Being that can be understood is language] (1999, 474), taken in a certain way, amounts to a kind of "lingualism" or linguistic idealism. Seebohm argues that phenomenologically higher levels of understanding (the understanding in language) are built on animalic understanding and experience at the pre-predicative (or pre-linguistic) level. Seebohm also provides an account of the understanding and interpretation of non-linguistic items, monuments, and material traces of historical and pre-historical human cultures. This has become more important as historians and archaeologists focus more and more on material culture.

Fifthly, Seebohm and Gadamer agree that psychoanalysis and the critique of ideology do not provide the paradigmatic exemplar of the problem of interpretation. Though neither refer to it explicitly, the principle of charity is fundamental for hermeneutics. Ideology, neurotic rationalization, and irony do provide particular problems for a charitable and trusting hermeneutics. Gadamer considers the first two of these in his exchange with Habermas and in the essay, "Text and Interpretation."[3] Seebohm speaks to ideology critique and psychoanalysis in this last book.

---

[2] See the last sentence of 1999, 491: "Rather, what the tool of method does not achieve must—and really can—be achieved by a discipline of questioning and inquiring, a discipline that guarantees truth."

[3] See my consideration of irony and Gadamerian hermeneutics, in Dostal 2008.

4.2.2 Their Disagreement About the First Canon

As noted above, Seebohm argues on behalf of the first canon, the canon that requires that we distinguish the interpretation from the application. This distinction presupposes that we can interpret the text in its own context as separate and other than our own. This canon importantly for Seebohm makes philology possible as a science. Seebohm writes: "The first canon is the watershed separating philology as a discipline from philology as a science" (2015, 115). The last sentence of *Truth and Method* (1999, 491, cited above) makes the same distinction. There is an agreement here with regard to the concepts of method and discipline, but Seebohm is primarily concerned to explicate the method and Gadamer, the discipline. In principle, Gadamer points to a way in which discipline can account for method, but Gadamer does not provide this account. Seebohm, on the other hand, provides an account of discipline and of method.

Gadamer's way of accounting for method, I would suggest, relies as his way so often does, on Heidegger. *Truth and Method* borrows from and relies on both the Heidegger of *Being and Time* and the so-called later Heidegger for whom language is often the central theme: "*Die Sprache spricht*" ["Language speaks"]. In *Being and Time* Heidegger distinguishes between the "apophantic as" and the "hermeneutical as."[4] The former is propositional and derivative. This includes the discourse of science. It is derived somehow from the more primordial hermeneutical experience of understanding. Gadamer follows Heidegger here. But neither Heidegger nor Gadamer provides an account of this derivation nor do they attend to what is derived (though one can find occasional comments about the empirical sciences).

If one looks carefully at the revisions that Gadamer made to subsequent editions of *Truth and Method* in conjunction with the comments that Gadamer made retrospectively with regard to this work, we find that the changes to the text are not extensive and are not substantive. Most of the additions to the text are footnotes that refer to this later work. The most changes in the text proper concern the natural sciences. Gadamer confesses later that he was not happy with the treatment of science in *Truth and Method*.[5] He admits that the treatment of science there is not adequate. But he does not see this inadequacy as relevant to the principle theses of his hermeneutics. His primary concerns relevant to science are two: (1) how the success of science in modern society has led to scientism, and (2) the role of technology and bureaucracy in modern scientistic culture. Obviously neither of these concern, except indirectly, the epistemology or ontology of science.

The first canon is directly relevant to the epistemology and ontology of science. Gadamer, as we have noted, rejects this canon. He speaks to it in a variety of ways.

[4] See Section 33 of *Being and Time*: "Statement [or "Assertion"—*Aussage*] as a Derivative Mode of Interpretation."

[5] See his comments in his autobiographical essay (1997, 40–41). See also the added footnote reference to Thomas Kuhn (1999, 283). I discuss the changes and additions to *Truth and Method* and the question of Gadamer's development in Dostal (2003).

He writes, for example, that "the effect of a living tradition and the effect of historical study must constitute a unity of effect" (1999, 282). He calls the rendering of the historical object of study as separate from the context of the interpreter an "abstraction" (282) and a "liberal fiction" (283).

Seebohm, on the other hand, defends the canon and the distinction between the contexts of the interpreter and the interpreted. He also defends the related distinction between interpretation and application. As we have noted, for Seebohm these distinctions make philology possible as a science. Seebohm is happy to call the distinction an "abstraction." What is called for is "disinterest" (2015, 152). This disinterested abstraction is closely related to the important phenomenological methodical principle of reduction (which we find Seebohm discussing and defending in this work). Seebohm writes: "To separate them [interpretation and application] requires an abstractive reduction" (56). This disinterest is a complicated and multi-layered one. It would be too much to follow Seebohm's account in detail here, but we should notice several important aspects of the account.

First, Seebohm acknowledges that:

> It is not possible to apply the first canon of philological hermeneutics requiring the separation of the context of the interpreter and the context of the interpreted texts or life expressions in general to the general structures of intersubjective temporality and elementary understanding in historical research and philological-historical research. (129)

This is to say that the necessary reconstruction of what is talked about in an historical text or what is found at a historical site is a reconstruction, as Seebohm says, "in the *present of the interpreter*" (129). The "elementary" aspects of life and the elementary understanding of these aspects are necessarily part of our context such that we cannot distance ourselves from them. This means that with regard to the elementary aspects of life and understanding, the distinction between understanding and application does not hold for Seebohm. But he argues that the canon does apply to first-order higher understanding in a past real cultural lifeworld (129).

Secondly, Seebohm acknowledges that the separation of contexts that is required by the first canon is not possible for contemporary history because, obviously, the context of the historian is the same context as that of the object of study. They share the same context.

Thirdly, Seebohm points out that in classical antiquity philology and rhetoric were identified as closely related aspects of the interpretive process. Their identification rendered interpretation and application inseparable. It also makes ancient philology epistemologically unscientific. Seebohm calls such a cultural tradition "archaic" (83)—elsewhere he calls it "humanist." Seebohm does not explicitly say it, but it would make sense for him to point out that Gadamer's project, among other things, is to rehabilitate certain basic features of ancient thought and his rejection of the first canon could be seen as important to this rehabilitation. From Seebohm's perspective Gadamer's hermeneutics can be seen as "archaic." There is a kind of objectification at the heart of the modern scientific method that Gadamer resists for his hermeneutics. Gadamer acknowledges in *Truth and Method* that objectification is at the heart of the modern scientific method. Here Seebohm and Gadamer agree.

For philology (hermeneutics) to attain the status of "science" it must take a distance from its object of study; it must objectify.

Fourthly, the distinction of interpretation and application entails a distinction between theory and practice. The abstractive reduction for the historical understanding of a text from a past culture requires that the scientific investigator suspend any practical interest in what the text claims or urges—the values, motives, purposes of which the text speaks. But Seebohm shows that we should distinguish theoretical interests from practical interests (293–294). Though practical interests are suspended, theoretical interests are not. The researcher is interested in determining what the values, motives, and purposes were without making a judgment about whether those values should be accepted or not. The researcher is interested in finding the relevant texts, in establishing the relevant context, and in attaining and making use of the tools necessary for her or his research. "Disinterest," a Kantian concept, is not the same as "uninterested." It is, rather, as I just indicated, a suspension of practical interest. The practice of the philological-historical method is motivated by theoretical interests.

Some commentators on Gadamer's hermeneutics think that Gadamer rejects the distinction between the theoretical and the practical, especially with regard to hermeneutics. Gadamer famously refers to his philosophical hermeneutics as "practical philosophy." But in the very essay entitled "Hermeneutics as Practical Philosophy" in which Gadamer argues that hermeneutics should be considered a part of practical philosophy, he concludes by writing:

> So when I speak about hermeneutics here, it is theory…. Hermeneutics has to do with a theoretical attitude toward the practice of interpretation…. This theoretic stance only makes us aware reflectively of what is performatively at play in the practical experience of understanding. (Gadamer 2007, 245)

Seebohm could accept this statement. In this essay and elsewhere Gadamer comments that theory, too, is just a form of practice (and cites Aristotle). This, too, fits Seebohm's account. Seebohm's discussion of the interests of theory concern the practice of theory, the practice of science. Neither Gadamer nor Seebohm should be taken to be a pragmatist, but there are some affinities.

Fifthly, and finally, Seebohm argues that the first canon does not hold in jurisprudence and theology—the two domains that Gadamer makes exemplary of the hermeneutical situation. Jurisprudence is directly concerned with how the law applies in a certain situation. Theology is, at least in part (moral theology), concerned with how God's "word" applies in our lived situation. We could say that Seebohm and Gadamer agree about this. But we should also recognize that Seebohm is critical of Gadamer for privileging these two domains as exemplary of the task of hermeneutics. In addition, we should also note that for Seebohm any historical study of law requires the invocation of this first canon.

Seebohm is keen on what is required for the scientific understanding of the interpretation of texts (and traces) from which the investigator/interpreter has some historical distance. In the essay by Gadamer just referred to, "Hermeneutics as Practical Philosophy," Gadamer provides a sketchy history of hermeneutics, a history that agrees with the much richer and fuller account provided by Seebohm. Most crucially, Gadamer points out (as does Seebohm) that with modernity comes the recognition of historical distance from the texts of classical antiquity. This recognition enabled the scientific philological approach of the nineteenth century—the approach that Seebohm is both criticizing and developing.

There is another nuance (the last to be considered here) with regard to this distance that time and history provide. Gadamer usually looks to the continuity over time and history. He often speaks about tradition as though it were simply continuous. Seebohm counters Gadamer by pointing out the importance of discontinuity in historical development and tradition. Seebohm likes particularly to point to the break between the pagan period and the Christian medieval world and also to the break between that medieval world and modernity that is largely affected by the Reformation. The Christian is not about to take the truth claims of the ancient writers about the Olympian deities seriously. The reformer is not about to take the truth claims of the medieval church seriously. The injunctions of the Greek poets and dramatists to honor the gods is not an injunction that the Christian will apply. The break in the tradition helps provide a kind of "natural" abstractive reduction whereby we are not in the first place concerned about the truth claims of the text.

There are several aspects of Seebohm's and Gadamer's respective treatments of tradition that could be unpacked to show that their differences here are not so great. But I will point to one very important and basic agreement in this regard. Though Seebohm may in certain contexts valorize discontinuity, he and Gadamer would agree that however much we find discontinuity and otherness and difficult obscurities in history and the tradition, what we find there should not in principle be considered incommensurable. Both Gadamer and Seebohm are committed to the commensurability of fixed life expressions. And commensurability does not mean that the other is, in the end, not other. It means, rather, that there is always the possibility of communication and understanding.

This leads me to a last comment about the distance that Seebohm sees as required for scientific philology and the human sciences in general. Were we to analyze carefully the conception of "dialogue" that Seebohm brings to his discussion of the human sciences, we would find that he uses the term in very much the same sense as can be found in Gadamer. "Dialogue" always seeks agreement and a fusion of horizons (2015, 295, 405). Seebohm argues that the appropriately understood methods of the human sciences require that the researcher/scientist communicate with the subject of study but not dialogue. Dialogue is to be excluded. There needs to be intersubjective communication so that the researcher can hear or read how the subject of study understands what she is doing. Dialogue goes beyond communicating what the subject is thinking, doing, feeling, and so on to attempt to persuade the researcher/scientist that she should agree with how the subject of the study understands and acts in the world—accept the truth and valuation claims of the agent under consideration. So while Gadamer asks us to dialogue with the text and to dialogue with the other person, Seebohm requires that we forego dialogue. This is another aspect of the implications of the first canon.

4.2.3 Why Their Hermeneutics Differ

In short, though it is a complex and nuanced matter as I have tried briefly to show, the primary difference methodologically between the hermeneutics of Seebohm and Gadamer is the status of the first canon. It is fair to say that Seebohm embraces it and Gadamer rejects it. But we have just noted that in two important contexts—contemporary history and those contexts when we look to a text to address what we should do (jurisprudence and moral theology)—the canon, for Seebohm, does not hold.

So how do we locate the difference between these two accounts of hermeneutics? The difference can be articulated simply. Gadamer's concerns are ontological and Seebohm's epistemological. This is how Seebohm formulates their differences and it seems that Gadamer accepted this. Gadamer recognized that there is "method" in philology and that the hermeneutical practice that he is describing presupposed good philology—that is, that one has the authentic text and one knows who the author is (not some forgery), and one knows what other texts the text with which one is concerned is responding to, and so on. Gadamer liked to point out that he was trained as a philologist and knew good philology from bad.

To use Plato as an example: Gadamer is concerned to show that in reading Plato we are to allow the text to make a claim on us, ultimately such that it might transform our lives. In other words, to understand Plato is at the same time to see how it applies to one's life. Seebohm's question in a certain sense is prior. How does the reader know the difference between an authentic Platonic text and a fake one? And within the authenticated texts, how do we sort out the variations in the text among various editions of the texts? Why does Gadamer accept the seventh letter as authentically Platonic when others do not? The answers to these questions (with various degrees of confidence) are provided by the philological-historical method. Gadamer's hermeneutics cannot account for them but, yet, in an important way, it presupposes them.

But what does it mean to say that Gadamer's concerns are ontological and Seebohm's epistemological? Seebohm tells us that his scientific philological methodology is concerned with validity and not with truth. Gadamer is concerned with truth. What does this mean? One way that Seebohm articulates this is to distinguish first order claims and the second order interpretation of the claims. First order claims are truth claims. The second order interpretation of these claims "brackets" the truth claims. Allowing the text to make a claim on us and to dialogue with the text, basic aspects of Gadamerian hermeneutics, is to remain at the level of the first order claims and to remain concerned with truth. Seebohm's hermeneutics takes a step back, brackets the truth claims, and works to achieve an intersubjectively valid interpretation of the first order speech with its truth claims.

Understood in this way, Seebohm's methodological hermeneutics and Gadamer's philosophical hermeneutics do not contradict one another since they are operating on different levels with different concerns. On the one level the first canon is not relevant; on the other level it is indispensable.

Though Seebohm does not present his hermeneutics in opposition to Gadamer's hermeneutics, he does take issue with some of Gadamer's interpretations of the work of the nineteenth century hermeneuts/philologists. From Seebohm's perspective Gadamer too much tries to paint them with the colors of positivism and Cartesianism. This is indicative of the fact that the rhetoric of Gadamer often presents Gadamerian hermeneutics in opposition to Cartesianism, positivism, and scientism in a way that suggests one should either accept his account of hermeneutical experience or one is left to embrace positivism. If one looks carefully, we can see that Gadamer does leave a space for methodological hermeneutics but often he neglects it entirely. If and when this space is neglected, we find ourselves in the position that Seebohm calls "hermeneuticism"—a "radical historism… beyond any interest in the possibility of objective validity in philological … research" (384). Seebohm is careful not to accuse Gadamer of this, but a philosopher (not mentioned by Seebohm) who embraced such a view would be Rorty. Seebohm's hermeneutics provide a third way that is neither "hermeneuticist" nor positivist.

However much Seebohm insists that his hermeneutics do not conflict with the ontological hermeneutics of Heidegger and Gadamer, he clearly does not wish to embrace Heideggerian ontology. But neither does Seebohm approve of what Husserl sometimes seems to be arguing for—a final foundation in an absolute or primal ego [*eine Letztbegründung in einem Ur-Ich*]. In the concluding two sections of this last book Seebohm argues quite clearly that any "metaphysical" grounding for phenomenology is irrelevant to phenomenological epistemology (see 390–413). But his comments on what he sees to be the metaphysical speculation of the late Husserl are not so clear and direct. He writes how the New School phenomenologists and the phenomenologists of Mainz did not follow this late "turn" by Husserl. Seebohm was affiliated both at the New School and at Mainz. Yet he utilizes third person pronouns to speak to this refusal: "they" did not follow the turn. Seebohm suggests that Fink influenced Husserl and that Fink, in turn, was influenced by Hegel and Heidegger. Seebohm wonders whether we might yet find something in Husserl's unpublished manuscripts that would resolve the incompatibility between a phenomenology that recognized that the lifeworld is the region of ultimate grounding and a phenomenology that seeks a grounding metaphysically beyond that (391). This question goes beyond the concern for an appropriate hermeneutics in the context of the historical sciences that was the primary concern for Seebohm in his last two books. And it goes beyond the question for this paper—a consideration of the hermeneutics of Thomas Seebohm.

References

Dostal, Robert. 2003. The development of Gadamer's thought. *The Journal of the British Society for Phenomenology* 34: 247–264.

———. 2008. Gadamerian hermeneutics and irony: Between Strauss and Derrida. *Research in Phenomenology* 38: 247–269.

Gadamer, Hans-Georg. 1997. Reflections on my philosophical journey. . In *The philosophy of Hans-Georg Gadamer*, ed. Lewis Hahn, Library of Living Philosophers XXIV. Chicago: Open Court.
———. 1999. *Truth and method*, 2nd rev ed. New York: Continuum.
———. 2007. *The Gadamer reader*. Ed. and Trans. Robert Palmer. Evanston: Northwestern University Press.
Seebohm, Thomas M. 1972. *Zur Kritik der hermeneutischen Vernunft*. Bonn: Bouvier.
———. 1977. *Ratio und Charisma. Ansätze zur Ausbildung eines philosophischen und wissenschaftlichen Weltverständnis im Moskauer Rußland*. Bonn: Bouvier.
———. 2004. *Hermeneutics: Method and methodology*. Dordrecht: Kluwer.
———. 2015. *History as a science and the system of the sciences: Phenomenological investigations*. New York: Springer.

Robert Dostal is the editor of the *Cambridge Companion to Gadamer* (2002) and author of numerous articles on Heidegger, Gadamer, Kant, and Arendt and on hermeneutics and phenomenology more generally. He has been a Humboldt Fellow at the Universities of Cologne and Freiburg. He completed his dissertation in 1977 at the Pennsylvania State University under the supervision of Thomas M. Seebohm.

Chapter 5
The Tasks and Contexts of Understanding in Dilthey and Seebohm

Rudolf A. Makkreel

Abstract One of the main tasks of Thomas Seebohm's *History as a Science and the System of the Sciences* is to properly characterize the human sciences and relate them to the natural sciences. Since one of his points of reference is the way Wilhelm Dilthey distinguishes those two kinds of sciences, I will show that Seebohm goes too far when he claims that Dilthey excludes explanative laws from the human sciences. That Dilthey merely delimits the scope of explanations to special contexts delineated by understanding will become clear when I expand on the interesting ways that Seebohm modifies Dilthey's distinction between elementary and higher understanding. Seebohm claims that if we define history as a fact-based science rooted in the natural attitude of the lifeworld then elementary understanding can already provide causal explanations. Because Dilthey has a more encompassing conception of history he doubts that elementary understanding can already explain its complexities. Only delimited socio-cultural historical contexts defined by higher understanding can set the stage for explanations.

Thomas Seebohm's *History as a Science and the System of the Sciences* is an ambitious work and demonstrates an impressive range of knowledge of the history and development of the sciences. One of its main concerns is to properly conceive the human sciences and relate them to the natural sciences. It goes back to nineteenth century efforts to establish history as a proper science and touches on the hermeneutical writings of Friedrich Schleiermacher and the philological methods of Johann Gustav Droysen. This leads to Wilhelm Dilthey's project to define the human sciences, which becomes the point of departure for Seebohm's efforts to give both the natural and human sciences a better epistemological profile by relating them to the contributions of Edmund Husserl's phenomenology and the developments made

R. A. Makkreel (✉)
Emory University, Atlanta, GA, USA
e-mail: philrm@emory.edu

© Springer Nature Switzerland AG 2020

T. Nenon (ed.), *Thomas Seebohm on the Foundations of the Sciences*, Contributions to Phenomenology 105,
https://doi.org/10.1007/978-3-030-23661-8_5

possible by Alfred Schutz's phenomenological reflections on the structures of the social world. The result is a very comprehensive treatment of the system of all the sciences.

No single essay can do justice to the many positions developed in this extensive work. I was asked to comment mainly on Seebohm's relation to Dilthey's theory of the sciences and his philosophical hermeneutics, and as I attempt to fulfill this task, I will also offer my own assessment of some of the interesting distinctions and controversial conclusions that Seebohm arrives at about the role of understanding in the human sciences.

5.1 The Relation Between Explanation and Understanding

Seebohm points out that, according to Dilthey, the natural sciences and human sciences have "different goals. The aim of the natural sciences is to *explain*, the aim of the human sciences is to *understand*" (2015, 71). This contrast highlights the main concerns of these two kinds of sciences, but for Dilthey this does not require as "strict [a] separation" (70) as Seebohm claims. Dilthey acknowledges that the natural and human sciences sometimes overlap, and in the first book of the *Introduction to the Human Sciences* he argues for a mere relative independence of the human sciences. The main task of a human science such as history is not to provide law-based causal explanations of the human actions and interactions that shape the socio-historical world, but to understand the meaning and value of what has happened. For Dilthey understanding the significance of human historical events requires being able to organize them in their proper contexts and to articulate the structural uniformities that can be found in this way. This contrast leads neo-Kantians Wilhelm Windelband and Heinrich Rickert to go further and propose a distinction between the natural sciences as being nomothetic in appealing to universal laws of nature and the cultural sciences as idiographic and interested in the description of individuality. Although Dilthey and the neo-Kantians did influence each other, Seebohm seems to think that Dilthey accepted their nomothetic-idiographic distinction. In fact, Dilthey explicitly rejected this neo-Kantian distinction. The understanding and appreciation of what is individual and distinctive is important in the human sciences, but Dilthey argues that the recognition of individuality can only be achieved against a general background. He regards all the human sciences as systematic in nature in that they are characterized by "the connection of the general and the individual" (2010, 227). Initially, this connectedness of the general and the individual in human life is more likely to disclose itself in organizational or structural ways rather than in terms of explanative laws. But Dilthey makes it clear that ultimately the human sciences must also "search for the causal relations that condition this individuation, the gradations, affinities, and types of human-historical life" (227). Lawful causal explanations may be harder to arrive at in the human sciences, but they are to be sought despite what Seebohm claims about Dilthey.

Let us start by considering what Dilthey says about the process of understanding [*Verstehen*] and its function in relation to the human sciences. His most famous contrast between explanation and understanding is to be found in his "Ideas for a Descriptive and Analytic Psychology," where he writes that "we explain through purely intellectual processes, but we understand through the cooperation of all the powers of the mind activated by apprehension. In understanding we proceed from the context of the whole… to then make the particular intelligible to ourselves" (147). Dilthey's main claim is that a descriptive or phenomenological psychology should not start by appealing to the kind of hypothetical and atomistic elements that ground traditional associationist psychology. It is possible to describe the stream of consciousness holistically as a structural nexus without looking for explanative laws of association to combine so-called discrete sense data. To be sure, the initial non-hypothetical connectedness that we find in experience is at times indeterminate, especially at the edges, and it is for this reason that Dilthey accepts "the significance that the consideration of hypotheses has for enlarging the horizon of descriptive and analytic psychology" (181). Causal hypotheses are deemed unnecessary for our initial understanding, but as we begin to analyze what has been described in general terms we come upon points of detail where causal explanations may be able to fill in gaps in our understanding. Although Dilthey worked to reshape psychology as a descriptive human science, he always took the experimental results of more naturalistic psychologists into account. A descriptive approach to experience allows us to understand the overall coherence of our mental life, but causal explanations can be used to fill in many of its details. Understanding thus provides the framework for determining where causal conditions become relevant. The consequence of this approach is to delimit the scope of causal explanations in the human sciences, but not to exclude them.

This kind of approach rooted in Dilthey's descriptive psychology, which the late Husserl recognized as a genial anticipation of phenomenology, also characterizes Dilthey's approach to history. Dilthey is skeptical about the existence of laws of overall historical progress, but open to more limited developmental laws that apply to the specific socio-cultural systems in which historical action takes place. He rejected Marx's general speculations about history at large but regarded some of his specific economic laws regarding the accumulation of capital in the nineteenth century as plausible (see Dilthey 1989, 40, where I cite [1991, 186–187]).

As was indicated earlier, in book 1 of the *Introduction to the Human Sciences,* Dilthey only argues for a relative independence of the human sciences. Thus he acknowledges that "Man, because of his position in the causal system of nature, is conditioned by it in a two-fold respect" (1989, 69). Human beings are obviously conditioned by natural stimuli through their senses and nervous system, but this need not fully determine their capacity to decide how to act. However, the extent to which we can freely decide to do something in accordance with our own purpose exposes us to a second mode of dependence on nature. If our efforts to execute a decision are to be successful, we must take the laws of nature into account. We cannot build a stable dwelling without taking the laws of gravity, etc. into account. But to admit that we are empirically dependent on nature on the one hand and on the results of the natural sciences on the other, does not mean that we are constrained by a naturalistic epistemology. Seebohm is partially correct in saying that Dilthey did not work out a new epistemology of the human sciences, but in fact Dilthey thought he did something better in his Breslau Draft for book 4 of the *Introduction to the Human Sciences.* There he supplements *Erkenntnistheorie* as an epistemological theory of cognition with an account of self-reflection [*Selbstbesinnung*] that provides "the foundation for action as well as for thought" (268). From Dilthey's final writings, it becomes clear that self-reflection is a mode of knowing [*Wissen*] that encompasses what the sciences allow us to cognize [*erkennen*] conceptually (2002, 24–27). Knowing as a mode of self-reflection critically assesses experience by contextualizing it and serves to "lay bare the foundation which supports the edifice of the sciences" (1989, 268). And from that perspective, the human sciences and their concern to understand the "total content and nexus of the facts of consciousness" (268) are more comprehensive than the natural sciences, which in Dilthey's words "explain only part of the contents of external reality" (203). What he calls the "intelligible world" of the mechanistic system of the natural sciences is a "highly artificial [cognitive] abstraction from what is given in experience… ." (268).

Seebohm does not take into account this Diltheyan way of distinguishing understanding as a mode of knowing from explanation as a mode of cognition. Thus while he does take note of Dilthey's distinction between elementary understanding and higher understanding that grows out of this, he develops it in ways that are quite different. An altered hierarchy of understanding is established by Seebohm that proceeds from animalic understanding to first and second order elementary understanding, to first and second order higher understanding culminating with the purposes of the social sciences.

5.2 Levels of Understanding

Animalic understanding is defined by Seebohm as involving an exchange in which the "life expressions of the other living body trigger as actions of the other living body immediate reactions of one's own living body and vice versa" (2015, 47). This exchange takes place in a natural lifeworld and turns out to also encompass our reactions to inanimate bodies by analogy. Animalic understanding is said to occur at the level of what Husserl calls "passive synthesis" (47) and would have to be characterized as a kind of behavioral awareness that lacks the requisites of understanding in Dilthey's sense.

Seebohm asserts that the next level involves elementary understanding: it moves beyond instinctive actions and reactions to make room for conscious interactions with other animate bodies as fellow participants in practical projects. At the same time, mere inanimate bodies come to be seen as the "raw materials" for practical ends and are thematized as part of the "natural environment in a lifeworld" (49). We now have objects that can be seen to be part of the natural world of the sciences as well as tools for a practical world. Seebohm's elementary understanding also takes account of the fixed life expressions of the historical past beyond the immediate life expressions of animalic awareness.

This characterization of elementary understanding gets closer to what Dilthey, the philosopher of history, was concerned with. But Seebohm conceives it mainly as a sphere of natural objects about which multiple subjects have competing practical aims. Understanding is pragmatically circumscribed in such a way as to abstract from its normative tasks and reduce it to a neutral kind of philological exegesis. By contrast, Dilthey relates elementary understanding to the medium of commonality—a localized form of Hegel's "objective spirit"—which endows objects with more than practical meaning. Dilthey's sphere of elementary understanding includes the conventions, customs, assumptions, and prejudices that we assimilate from our regional community. Elementary understanding encompasses what we have inherited from our local past and orients us as we live our life. It is not just geared to practical results, but also includes the broader social and cultural dimensions of our experience. For Dilthey elementary understanding provides a quite comprehensive mode of acquaintance with what is shared in one's community. But eventually this kind of know-how (we could call it "life knowledge") runs into unanticipated problems. It is when some of our normal expectations are frustrated that we need to turn to higher understanding where intellectual analysis and cognitive methods are brought to bear. Elementary understanding for Dilthey is a pre-scientific, but rich and multivalued mode of immediate knowing, which the higher understanding of the human sciences then subjects to cognitive disciplinary analysis and refinement. But because higher understanding tends to resolve specific problems, its results need to also be reflectively assessed to determine whether they do justice to our overall outlook. This involves a circular hermeneutical process that guides historical interpretation from elementary understanding to an intermediate higher cognitive understanding, to a final reflective knowing that evaluates how our initial assumptions and beliefs measure up.

As already suggested, Seebohm's conception of the tasks of elementary and higher understanding and their relation to history is somewhat different. History is set apart from the other human sciences by Seebohm and made into a basically fact-establishing philological discipline. It applies a methodologically abstract mode of factual inquiry that places what is to be understood in a naturalized lifeworld. Seebohm defends this abstractive reduction of history as providing him a link between the natural and human sciences. This is what he writes:

> In its interpretations of the meaning of the actions, interactions, and events in the past for the past, history is still bound to the methodological abstraction of the philological-historical method, but history transcends the limits of this abstractive reduction because its reconstructions presuppose the spatial and temporal structures of the lifeworld in general. History accordingly includes dimensions of the lifeworld that also belong to the natural environment and this means the basis of the methodological foundations that are constitutive for the natural sciences. In this respect history has the potential of serving as a mediator in the alleged strict opposition between the natural sciences as sciences of explanation and the human sciences as sciences of understanding. In this respect the relation between history and the social human sciences is of crucial epistemological significance. (2015, 95)

History is made into a fact-based science rooted in the natural attitude of the life-world and is supposed to bridge the gap between explanation and understanding. But this comes with a cost. Seebohm has accepted Rickert's characterization of history as an idiographic science that confirms individual facts. However, as was already indicated, for Dilthey and such heirs as Hans-Georg Gadamer and Paul Ricoeur all the human sciences have systematic aims that consider facts in some meaning context. This is as true for psychology as for sociology or economics, but it is especially true for history. Dilthey's descriptive psychology deals with individual experience, but this experience is temporal and always framed by a historical context that orients us. Not only is there a past life course that influences the significance of present impressions, but there are also expectations about what is to come. The socio-cultural human sciences obviously refer to a more explicit public context. Finally, history is a nexus of events that can only be made sense of by focusing on those socio-cultural contexts or systems that frame and channel what is happening.

Seebohm also brings a context to bear in his discussion of the methodology of history. But this is a restricted, primarily natural context, not yet the overall context of life experience. This becomes clear from the way Seebohm bases his historical methodology on the first philological canon of hermeneutics, which treats its data as texts. He defines his first canon as follows:

> (1.a) The text ought not to be understood out of the context of the interpreter. It ought to be understood out of its own context and out of the context of the contemporary addressees. (2015, 112)

Let me clarify that when Seebohm says that a text must be understood solely "out of" the context of its contemporary addressees, he does not mean "outside of" that original context, but from within it. Thus the first canon does not allow subsequent and present interpreters to consider its relevance for them. They are excluded from adding to the original meaning context. Present interpreters must reconstruct what the text meant for the author and how the readers of his or her time and circumstances would have understood it. This limit on the interpreter parallels the one that Antonin Scalia's doctrine of originalism placed on judges in interpreting the U.S. Constitution. This is not to say that Seebohm would agree with Scalia, because he exempts jurisprudence from his first canon. Jurists have an overriding prudential or practical task that makes the application of the law to the present essential. Thus Seebohm writes: "Application is an inseparable part of the interpretation of law texts because the *meaning* of law texts *implies* that law texts have to be applied to cases given in *present* experience" (354).

As Robert Dostal points out in his essay in this volume, despite the great emphasis that Seebohm places on the theoretical importance of the first philological canon, he exempts not only jurisprudence, but also theology from this canon because they are by nature practical. Seebohm's critique of Gadamer is that he made these two disciplines exemplary for all the human sciences and thereby merges interpretation and application in all cases.

Seebohm is especially insistent that historical interpretation be assigned a primarily theoretical task that suspends the task of application. History, as he conceives it, is concerned with the interpretation of remnants of the past, whether these be archival documents and texts or other human artifacts. The theoretical task of history is to reconstruct the original intentional context within which an author addressed, or an agent affected, his contemporaries. Now to be sure, Seebohm is not merely invoking the mental intentions of the author or agent. His intentional context is geared mainly to natural facts, as can be shown on the basis of the following assertion in his text:

> "Historical fact" is a term for a genus of essentially different kinds of facts, i.e., events in the natural environment, actions, motives for actions, intended or not intended consequences of actions, beliefs, habits, customs, social and legal relations, and cults. (2015, 125)

Seebohm's aim here is to place human actions and their motives into a natural environment that can provide a stable base for further inquiry. It makes historians treat the past as a fixed remnant rather than as having a productive afterlife. What is interesting about Seebohm's account of history as a human science are the parallels that he makes with natural history. This allows him to speak of both as concerned with the understanding of traces. "Traces of the history of nature are (1) geological traces and (2) fossils. The traces of history as a human science are (3) fixed life expressions" (260). The understanding of natural traces is said to be of a first-order kind, but the understanding of human traces is of a second-order kind that acknowledges that a first-order understanding is already encoded in life expressions. But qua trace, even a life expression is said to be capable of being explained in terms of its place in the naturalized lifeworld.

Now Dilthey would not contest Seebohm's appeal to a "natural environment" as a background. But he would not make it the defining framework. Our lived environment is as much social as natural. Our natural environment cannot be separated from the communal background of social and cultural institutions that frames human experience and interaction at all levels. Seebohm might seem to also include our socio-cultural environment when he speaks in the above quote of customs, social and legal relations, and cults. But all of these commonalities are treated as practical incentives that motivate individual human beings as they react to each other as animate beings. Seebohm's focus on natural practices comes out especially in the central facts listed above, namely, "motives for actions, intended or not intended consequences of actions, beliefs [and] habits" (125). The search for motives and beliefs is used to reconstruct facts about individuals in the past—facts that have a fixed meaning that is not subject to reinterpretation in light of the subsequent effective or productive history [*Wirkungsgeschichte*] of what they have done. To be fair to Seebohm, he later turns to a second hermeneutical canon that allows historical data to be viewed in relation to larger wholes that include the present interpreter. He credits Dilthey with generalizing this second hermeneutical canon into a "meta-canon for methodical rules of the human sciences in general" (103) and the basis of the hermeneutic circle according to which the parts and wholes of what is given are interdependent. However, Seebohm will not put this second canon into practice without phenomenological clarification nor without the help of the second-order higher understanding made possible by the social sciences.

Seebohm uses the first canon to secure a historically reconstructed, factual base for his theory of the sciences. He seems to think that Dilthey, too, was a kind of reconstructionist, and this certainly was Gadamer's view of Dilthey. To be sure, in his *Poetics* of 1887, Dilthey did speak of reproducing the state of mind of the poet. However, he came to recognize that this could encourage psychologism and therefore becomes much more careful later when he speaks of re-experiencing [*nacherleben*] the work of a poet. Dilthey claims in the *Formation of the Historical World* of 1910 that re-experiencing a poetic work does not give us a representation of "the inner processes in the poet ... rather [it presents] a nexus created in them but separable from them. The nexus of a drama consists in a distinctive relation of material, poetic mood, motif, plot and means of presentation. Each of these moments performs a function in the structure of the work" (2002, 107 [1927, 85]). This separation of a work from its author allows Dilthey to aim for a better understanding rather than an identical understanding. What Dilthey would especially object to in the above quote from Seebohm about human actions in a natural environment is his appeal to the motives and beliefs of past historical actors and creators. Since Dilthey made room for psychology in his efforts to understand human life, this suspicion of motives and beliefs might seem surprising. But his psychology looks for how description can illuminate the developing nexus of consciousness. Since motives can have deep instinctive sources that are often unconscious, and beliefs are often quite private and ill-defined, appeals to them are considered by Dilthey to be highly hypothetical and encourage speculative kinds of explanation that cannot be adequately tested. He writes that "historical skepticism can be overcome only if our methods are not expected to ascertain motives… [and] when psychological subtlety is replaced by the understanding of the products of the human spirit" (279).

Seebohm is convinced that reconstructive historical understanding provides facts that can be made sense of in a naturalized lifeworld geared to the practical interest of individuals and that this makes them readily available to being causally explained. Dilthey by contrast begins his descriptions of human experience and action without assuming that everything needs to be naturalized and causally explained. The purpose of history is not just to explain why and how things happened, but to frame these events in ways that allow us to understand their significance and their consequences. Elementary understanding provides an initial, common sense way of surveying how things are ordered and organized in human life, but it should stay clear of Seebohm's "elementary understanding of causal connections" (2015, 129), which will remain quite hypothetical without the application of higher understanding. Premature efforts to find causal explanations for complex human interactions introduce a hypothetical element into history that Dilthey would find worrisome.

Whereas Seebohm naturalizes elementary understanding in order to find a scientific base for history, we saw that Dilthey is quite willing to consider elementary understanding as pre-scientific. By expanding on Dilthey and alluding to Goethe, we could say that elementary understanding is rooted in what we *passively inherit* [*ererben*] from the common context in which we grow up.[1] It is higher understanding that then brings the universal standards of the human and natural sciences to bear on how human life is organized in order to explicate or articulate structural patterns. The regional commonalities of elementary understanding are compared and tested for their possible universal import. Certain aspects of more generally shared organizational structures may eventually disclose causal dependencies. These results of higher understanding we can only *actively acquire* [*erwerben*] through the cognitive methods of the human sciences. While the human sciences aim at universal cognitive results, they end up fragmenting our understanding of the world through a plurality of conceptual discourses. That is why it becomes necessary to move on to a kind of *reflective knowing* that is both theoretical and practical and that seeks to integrate what has been fragmented by scientific cognition.[2] This third level does not just rest upon the lower levels, for it returns to the first level as a re-experiencing that re-orients initial understanding. Although Dilthey speaks in traditional terms of grounding the human sciences, his approach is orientational along the way.[3] The foundation that Dilthey offers does not come at the beginning and only begins to dawn on us at the third level of reflective knowing. Only then do we begin to *possess* [*besitzen*] the foundation as our own.

5.3 What Kind of Foundation Can the Sciences Offer?

In that both Dilthey and Seebohm speak of elementary and higher understanding one could say that foundationalism applies to both, but in different senses. Dilthey's elementary understanding allows for the intersection of more factors from the start. Seebohm begins with a much more limited base and stratifies things more sharply as he moves up. He admits that his elementary system of goal-directed interactions

[1] "Was du ererbt von deinen Vätern hast, /Erwirb es, um es zu besitzen!" Goethe, *Faust*, lines 682–63.

[2] What I call "reflective knowing" relates how Dilthey describes self-reflection at 1989, 268, and knowledge [*Wissen*] at 2002, 25. In line 3 of 1989, 268, "self-reflection" and the "theory of knowledge" are incorrectly contrasted. When that translation was published in 1989, I did not yet realize that for Dilthey cognition is only one aspect of knowledge. Self-reflection is really being contrasted to a narrower *Erkenntnistheorie* as a purely intellectual theory of cognition.

[3] In *Orientation and Judgment in Hermeneutics* (2015), I develop a more explicit orientational approach to show how the interpretation of history must come to terms with the customs, conventions, and beliefs that we *assimilate* from the past. Our first relation to the past is how we are affected by it in elementary understanding, not by some philological representation of historical facts. After coming to terms with what is assimilated, we can *acquire* cognitive clarity about the past, and for this we have a variety of tools such as philology and other human sciences. But it is not productive to sharply differentiate these disciplines into purely theoretical ones such as philology and purely practical ones such as the social sciences. The ultimate task of historical understanding is to judgmentally *appropriate* all this reflectively as our own. See especially pp. 9, 81–84, and 194–97.

is "a partial dependent structure within the structure of a cultural lifeworld in general" (51). It provides a narrow base that claims to be factually secure and does not need to be reinterpreted. Higher understanding then gives us access to "the lifeworld as a whole and its essential aspects" (51). His higher understanding extends the scope of elementary understanding by focusing on "the social structures of the lifeworld … including customs such as fashions and styles, as well as customary laws and written laws with their distinctions between right and wrong in interactions" (51). These contents that Seebohm sees as being added by higher understanding were already contained in Dilthey's broader conception of elementary understanding. Seebohm's focus on "the needs of the practical activities of elementary understanding" (51) overlooks the way elementary understanding is already influenced by the customs and the institutions of objective spirit. To be sure, higher understanding is needed to analyze these structures and look for more general uniformities. Elementary understanding may start out being more provincial in scope in that it relies on common sense rather than the disciplinary methods of the sciences, but nothing is ruled out-of-bounds for elementary understanding. The "cults, artefacts serving cult activities, myths, prophetic revelations and poetry" that Seebohm defines as "manifestations of higher understanding" are already the subject matter of Dilthey's elementary understanding. Seebohm calls elementary understanding "blind without higher understanding" (52). I would call it less far-sighted than higher understanding.

The next step in Seebohm's account of higher understanding is most interesting. He again distinguishes between first-order and second-order modes of understanding. For elementary understanding, the difference concerned the subject matter: first-order understanding was about natural traces and second-order understanding about human traces. For higher understanding, even the first-order kind is mostly about human traces. The difference now is how human traces or life expressions are to be framed or contextualized. For higher understanding the first-order agenda is to place human traces in the naturalized environment of "the present lifeworld," whereas second-order higher understanding is about "written discourses representing the tradition of a present cultural lifeworld" (56). This suggests a shift from understanding things in terms of a Husserlian world to interpreting them as part of a more Gadamerian one. Seebohm writes that "first-order higher understanding is the *creative* understanding of the context of the lifeworld as a whole and of partial aspects … [such as] structures of political power, of custom and law, of trade relations … and the natural environment" (76). This kind of understanding is said to be found in "a prophet or a poet or a lawgiver or a natural scientist" (76–77). Even though natural scientists focus on explaining what happens in nature, their theories and discoveries contribute to an understanding of their lifeworld. It is not clear, however, what kind of understanding poets and natural scientists have in common. The creativity of poets is often not fully understood by them and draws on the unconscious. Whatever characterizes poetic insights, they do not have the cognitive status that is assigned to the theories of natural scientists. Lawgivers would certainly be aiming at a more rational world order that is comparable to that of natural science. Perhaps what prophets, poets, lawgivers, and natural scientists have in common is the desire to improve our life by changing either our relation to the world or the world itself. Seebohm explicitly leaves out human scientists from this group capable of first-order creative understanding.

Human scientists are assigned the task of second-order higher understanding, which is the "re-creative understanding of the manifestations of first-order creative understanding in immediate and fixed life expressions, the *interpretation* of life expressions" (76). The second-order higher understanding of the human sciences is said to appeal to psychology. Initially, this is an individual psychology, not unlike that of Dilthey. But a technique of diagnostic dialogue is added by which an individual pathology can be analyzed in conjunction with another individual. Seebohm's approach here is reminiscent of how Jürgen Habermas tried to redefine the logic of psychoanalysis as a "combination of hermeneutic understanding with causal explanation" (Habermas 1987, 270–273). Seebohm also discusses productive dialogues aimed at "repairing disturbed communication between participants" (2015, 405) and this leads him to move from individual psychology to social psychology. Acknowledging that we cannot understand the lived experience of other persons merely on the basis of individual psychology, Seebohm sees the need for a special social psychology that studies "the subjective reactions to the systems of social interactions of the subject's social environment" (406). Without this supplement, psychology is thought to remain interpretive and stop short of the scientific explanations made possible by the natural and social environment of individuals. Actually, Dilthey's descriptive psychology already includes second-order concepts that take our social context into account. Thus he points out that we cannot adequately understand a modern person's character trait of thrift without reference to the capitalist economic system. Dilthey defines human individuals as points of intersection of social and cultural systems. It is this insight as most fully developed in the *Formation of the Historical World in the Human Sciences* that leads him to stress that individuals are active participants in a variety of social, economic, and cultural systems. These systems may be small-scale enough so that individuals can see them as providing them the opportunity to extend their own influence. The differentiated nature of these socio-cultural systems works against Hegel's idealistic and deterministic conception of objective spirit that makes individuals subservient to nation-states. For Dilthey, each socio-cultural system can at best lay claim to only a part of an individual. Seebohm endorses this Diltheyan stance when he writes of individuals that "they are independent parts because they can be parts of different wholes of social interactions" (406). But beyond this point of convergence we see Seebohm and Dilthey diverge again when it comes to their conception of the social sciences. Seebohm writes that "research in the social sciences has to begin with the 'construction' of ideal types of the basic structures of social interactions and systems of social interactions" (407). These systems are constructed in terms of ideal purposes and geared to the prediction of future outcomes. Dilthey's theory of the human sciences also starts with an analysis of projected purposive systems, but he always plays them off against actually existing institutional forces and structures that work to resist or modify desired progress. In his final writings, Dilthey describes social structures that have productive outcomes that are not anticipated by projected purposive systems.

He always claimed that while the natural sciences may mathematically *construct* ideal systems, historians of the human world have the more difficult *formative* task of discerning the actual structures at work in the more human historical world: *Konstruktion* must be replaced by *Aufbau*. Ideally constructed systems abstract from many of the variables operative in reality: they resemble what is artificially simulated in the labs of chemists and physicists. To be sure, any attempt to understand the human world in its complexity cannot help but give more emphasis to some factors than to others. But the claim that we can ideally schematize social interactions to the extent that we can predict our human future makes Seebohm a kind of utopian. Thus he writes that "ideal types can serve as schemata for final and efficient causal explanations in social history … [making] "predictions in the present about future steps in the process of social interactions … possible" (407). A Diltheyan, too, thinks that knowing our past can help in anticipating at least some of the consequences of proposed actions, but to venture more specific predictions seems unrealistic.

5.4 Concluding Observations

I will end with two final considerations. First, it is important to make the case against a widely-held assumption that understanding tends to be a rather basic intuitive act that involves a kind of empathic identification with others. I do not think that Seebohm goes that far even though he allows for a merely animalic kind of understanding based on emotion that Dilthey would exclude from the process of understanding. Seebohm takes understanding more seriously when he speaks of the philological study of texts and other historical documents. He acknowledges that understanding can provide an important factual base for inquiry, but he regards it as merely a prelude to explanation. To be sure, understanding should serve as the starting point for explanation in that it provides the contextual conditions that set the stage for the complex problems that need to be resolved. Explanations can often provide specific resolutions. But as I indicated, understanding can instantiate a more comprehensive achievement than explanation. Understanding, especially the reflective kind, is a kind of knowing that calls on all our powers, not just the cognitive intellect. No intellectual explanation can replace such understanding; it can merely fill it in either to deepen our understanding or modify it. But it will still be a kind of understanding that we will end up with.

My second concluding observation concerns the nature of history and raises a question that has been intimated, but not fully spelled out. If, as Seebohm claims, the social sciences are present- and future-directed, what fills the gap between history as the reconstruction of an often distant past in accordance with the first canon on the one hand and the forward-looking perspective of the social sciences and their predictions on the other? The first canon is supposed to keep historical interpretation purely theoretical. The second canon makes room for its application to the present in light of future aims of the social sciences. But what is to prevent this practical application from becoming anachronistic and arbitrary? Let us consider what Seebohm says about jurisprudence. He distinguishes two kinds of roles for jurists: they have "to know the pre-given and presently enforced system of positive law" (347) as the background for the practice of jurisprudence, which aims at a prudential "creative act … guided by purposes and values" (413). But it is questionable whether this "immediate past horizon of the present," (412–413) which Seebohm appeals to as setting the stage for his social science ideals, provides enough of a historical context. What is also needed is a more extensive, intermediary sense of history as a "productive nexus" [*Wirkungszusammenhang*] that informs the present. Many historical achievements, whether they be economic, political, scientific, or artistic, attain an afterlife through institutionalization. Thus in considering what Marxism might still be able to contribute to our present age of economic dislocation, corporate corruption, and income inequality, it would indeed be important to re-read such classical sources as Marx, Engels, and Lenin and get them right. But the intervening histories of Soviet and Chinese communism and European forms of socialism are just as important in assessing the capacity of Marxism to deal with economic and social injustice.

History should neither be just a purely theoretical attempt at creating "simultaneity" with some stage of the past nor should it be made subservient to attempts at application to give it "de facto contemporaneity" (Gadamer 1999, 86). Gadamer, who distinguished these two alternatives for historical understanding, chose the contemporaneity option for reviving the Western tradition. Seebohm embraced both alternatives but assigned the former to history and the latter to the social sciences. But why not allow historians themselves to make what they have found to be true about some stage of the past relevant to the present by assessing its subsequent efficacy and its aftereffects? An adequate conception of history would have to reject each of the two Gadamerian alternatives accepted by Seebohm as too limiting. A first canon simultaneity-understanding and a second canon contemporaneous-application do not add up to a responsible historical interpretation. History as a human science is more encompassing. It is not just a factual idiographic resource that can then be mined for utilitarian purposes, for it has a direct bearing on the more general reflective and evaluative concerns that inform what it means to be human. For that bearing to disclose itself, history must be understood as a comprehensive continuum—however episodic—that we still belong to.

References

Dilthey, Wilhelm. 1927. *Der Aufbau der geschichtlichen Welt in den Geisteswissenschaften*, ed. B. Groethuysen. Göttingen: Vandenhoeck & Ruprecht.

———. 1989. *Selected works, volume I: Introduction to the human sciences*, ed. Rudolf A. Makkreel and Frithjof Rodi. Princeton: Princeton University Press.

———. 1991. *Zur Geistesgeschichte des 19. Jahrhunderts*, ed. Karlfried Gründer and Frithjof Rodi. Göttingen: Vandenhoeck & Ruprecht.

———. 2002. *Selected works, volume III: The formation of the historical world in the human sciences*, ed. Rudolf A. Makkreel and Frithjof Rodi. Princeton, NJ: Princeton University Press.

———. 2010. Contributions to the study of individuality. In *Selected works, volume II: Understanding the human world.* Princeton, NJ: Princeton University Press.

Gadamer, Hans-Georg. 1999. *Truth and method*. 2nd rev ed New York: Continuum.
Habermas, Jürgen. 1987. *Knowledge and human interests*. Boston: Polity Press.
Makkreel, Rudolf A. 2015. *Orientation and judgment in hermeneutics*. Chicago: University of Chicago Press.
Seebohm, Thomas M. 2015. *History as a science and the system of the sciences: Phenomenological investigations*. New York: Springer.

Rudolf A. Makkreel is Charles Howard Candler Professor Emeritus of Philosophy at Emory University and the author of *Orientation and Judgment in Hermeneutics* (2015, 2017) as well as of *Dilthey: Philosopher of the Human Studies* (1975, expanded 2nd ed. 1992) and *Imagination and Interpretation in Kant: The Hermeneutical Import of the "Critique of Judgment"* (1990, 1994). Also he is a co-editor of six volumes of Dilthey's *Selected Works*, *The Ethics of History*, *Neo-Kantianism in Contemporary Philosophy*, and *Recent Contributions to Dilthey's Philosophy of the Human Sciences*. He was editor of the *Journal of the History of Philosophy* from 1983 to 1998 and was awarded fellowships/grants by the NEH, DAAD, Alexander von Humboldt Foundation, Thyssen Stiftung, DFG, Volkswagen Stiftung, and the Heilbrun Fund. His works focus on German philosophy from Baumgarten and Kant and on hermeneutics, philosophy of history, and aesthetics.

Chapter 6
Phenomenological Reduction and Methodological Abstraction

Roberto J. Walton

Abstract Methodological abstractions, according to Thomas Seebohm, set up as a residuum the region of objects for a given science. The phenomenological reduction can be regarded as an abstraction whose residuum is the sum total of possible intentional objects as correlates of consciousness. Whereas a methodological abstraction in the broad sense of a phenomenological epistemology is concerned with all intentional acts and the evidence in which their objects are given, a methodological abstraction in a narrow sense deals with the methodology and objects of specific empirical sciences. Within the narrow sense, a first abstraction determines the ontological region of natural sciences by excluding all concerns except a theoretical interest. A further abstraction determines the region of the hard sciences and amounts to an abstractive reduction when the attempt is made to explain on this basis phenomena relevant to other sciences. As regards the phenomenological reduction, Seebohm analyzes the role of the egological and primordial reductions and highlights the genetic foundation of intersubjectivity within the structures of the primordial sphere. These structures include understanding whose basic form is a bodily understanding that functions as the genetic foundation for the elementary understanding of practical life, the first-order creative higher understanding inherent to cultural formations, and the second-order higher understanding provided by the interpretation of human sciences. A methodological abstraction determines the region of human sciences by bracketing practical moments of the lifeworld given in previous modes of understanding. After setting forth this overall framework, the article shows how elementary understanding plays a role in the generative foundation of the concepts both of natural and human sciences. Secondly, it examines the epistemic interpretation of the phenomenological reduction advocated by Seebohm in contrast to an ontic interpretation which is deemed irrelevant for a phenomenological epistemology. Finally, the article attempts to show that there is more to this elucidation if a deeper consideration is given to the pre-ego. It is argued that the epistemic interpretation and the ontic interpretation leave room for one another and even condition each other.

R. J. Walton (✉)
Universidad de Buenos Aires, Buenos Aires, Argentina

© Springer Nature Switzerland AG 2020

T. Nenon (ed.), *Thomas Seebohm on the Foundations of the Sciences*, Contributions to Phenomenology 105,
https://doi.org/10.1007/978-3-030-23661-8_6

The transcendental-phenomenological reduction entails an abstraction from the givenness of the world in the direct intention of the natural attitude in order to reflect an oblique intention on the objects and the world as correlates of intentional acts. Its abstractive reduction or reductive abstraction brackets the natural givenness of the world. In turn, the residuum of the transcendental reduction can undergo an egological reduction and a primordial reduction in order to show the manner in which intersubjectivity is implied in the constitution of transcendent objects. The residua of these further reductions point out abstract aspects within the residuum of the general phenomenological reduction. Hence, they are also a kind of reductive abstraction. These analyses provide a basis for a solution to the paradox of subjectivity formulated by Husserl in the *Crisis* (1970, §53). Seebohm also refers in a narrower sense to the methodological abstractions that are constitutive for empirical sciences because they determine as a residuum the limited ontological region of objects that make up their domain. Methodological abstractions in science have developed historically, and phenomenological epistemology attempts to reveal their generative foundations in the basic structures of the lifeworld. Once these main points about the phenomenological reduction and the paradox of subjectivity are outlined, with regard to the distinction between an epistemic and an ontic interpretation of the phenomenological reduction, it is argued that Husserl's two-fold approach to an ultimate ground in the primality of the ego has not been given adequate attention.

6.1 Methodological Abstractions in the Broad Sense

Seebohm offers the following characterization of the phenomenological reduction: "The reduction 'brackets,' i.e., it abstracts from the givenness of the world in direct intention of the natural attitude … "(2015, 15). Abstraction from givenness in this direct intention yields as a residuum, in the oblique intention of the phenomenological attitude, the givenness of the universal structure conveyed in the formula "ego cogito cogitatum." A residuum is an abstract moment that "cannot be given by itself," but "can be considered by itself only with the aid of an isolating abstraction bracketing the whole in which it is given" (2015, 18 n.). On account of this abstraction, Seebohm holds that "a universal phenomenological reflection is one-sidedly founded in the natural attitude" (1989, 377). What is bracketed in the phenomenological reduction is not the transcendence of the world, but rather "the claim that the world is the first evidence underlying all other evidence" (376; see 365f.). The residuum is made up by the objects and the world as the sum total of possible intentional objects in the way in which they exist as correlates of intentional acts. In the case of phenomenology, the region of objects is made up by the manifold manifestations of the structure "ego cogito cogitatum." The abstraction that sets aside the natural world is a way of putting into practice a procedure that Seebohm depicts with these words: "In the following investigations abstractions that determine in their residuum the region of objects that can be objects of a methodology of a science will be called methodological abstractions" (2015, 7).

In the narrower sense, a phenomenological epistemology or theory of knowledge deals with the methodologies of specific empirical sciences. In the broad sense, phenomenological epistemology is a general theory of knowledge concerned with all subjective intentional acts and intersubjective activities and with the evidence in which the objective correlates of intentional acts are given (40f., 87f.). In this broad sense, the phenomenological transcendental reduction should be included within the methodological abstractions. In other words, we can speak of a methodological abstraction in the narrower sense of specific empirical disciplines and a methodological abstraction in the broad sense of a general theory of knowledge.

On the basis of the reduction, phenomenological reflection in oblique intention undertakes the description of the general structures of consciousness. According to Husserl, as is well-known, in direct intention, objects are given with provisional evidence because they are open to counter-evidences and the world is given with open presumptive apodictic evidence because its givenness withdraws itself from negations or modifications of previous evidences. In contrast, consciousness is given to itself in oblique intention with apodictic evidence understood as "an evidence that is re-enacted in every act such that every act includes it in affirming, doubting or even negating it" (Seebohm 1989, 362). Every reflection on the existence of consciousness is performed by consciousness in a re-enactment of itself. An important point for Seebohm is that all acts of oblique intention presuppose either immediately or mediately an act of direct intention turned to objects and the world.

Two possible interpretations of the phenomenological reduction are examined. The epistemic interpretation is centered in the scope, the limits, and the evidences in which the objective correlates are given. And the ontic interpretation considers that the epistemic categories indicate an ontology concerned with different types of being (necessary, absolute, contingent) and also that the world as the totality of objects is the contingent correlate of the synthetic intentional acts of subjective consciousness as an absolute being. The focus of Seebohm's analyses is on the epistemic interpretation of the reduction as "the general methodological framework for a phenomenological epistemology" (2015, 40). The ontic interpretation is regarded as irrelevant for a phenomenological epistemology. In his article "Transcendental Phenomenology," Seebohm had avoided an ontological interpretation of the transcendental phenomenological reduction: "What is important is only the epistemic question about apodictic evidence, i.e., about the mode of givenness, rather than the ontological question about being" (1989, 364). He goes on to say that Husserl's project "has to be understood in epistemic terms, and in this framework the question of being is 'not admitted'," and adds that "the whole enterprise becomes absurd if it is interpreted within the framework of an ontology" (370).

An important point here is that two interpretations of the structure "ego cogito cogitatum" can be given. Either the three components can be understood equally as dependent parts of the whole encompassed by the formulation, or the ego can be interpreted as the active and unifying source of intentional activity that has the function of a necessary condition of experience (see Seebohm 2015, 35). This is the issue posed by the paradox of subjectivity, i.e., the paradox that "consciousness is given for itself as *being* the subject of the sum total of all objects in the reflective transcendental attitude, but it is also given to itself as *being* an object in the world in the natural attitude" (2015, 385; see also 34, 39f.).

According to the ontic interpretation of the paradox, the first horn of the dilemma implies that consciousness as the necessary being has the contingent being of the world as a correlate of its intentional activity. This entails an ultimate grounding in an absolute source of the correlation. The second part of the formulation states that consciousness is given with its correlative objects in oblique intention because it is given to itself as a contingent being among other contingent beings in the world. This interpretation for the paradox is given by Husserl in the *Cartesian Meditations* and the *Crisis*. In contrast, Seebohm works out an epistemic interpretation.

6.2 Abstraction in the Egologic and Primordial Reductions

If subjective consciousness has a necessary being and intentional objects are contingent beings, the problem of solipsism emerges and an explication of the givenness of the Other becomes necessary. Husserl attempts a solution to this problem accomplishing two further reductions within the phenomenological reduction: the egological and the primordial reductions. These additional reductions "are implied as correlates of abstractive reduction within the residuum of the universal transcendental-phenomenological reduction" (Seebohm 2015, 87). They render possible the task of showing the intentional acts in which Others, whether possible or actual, are given within the primordial sphere. Their task is "to determine the specific problems of the givenness of intersubjectivity and the Other as specific aspects within the phenomenological reduction" (42).

The egological reduction presupposes the givenness of Others within the residuum of the phenomenological reduction and determines how they are implied in the constitution of the objective correlates of subjective consciousness. It takes for granted the possible givenness of Others that experience, at the present time, all those sides of an object that, although I have seen them in the past and might experience them in the future, I cannot experience in the actual Now. Within the residuum of the phenomenological reduction, the egological reduction highlights the contribution of Others to the constitution of transcendent objects. It separates the transcendent object in the positive sense of objects given for an intersubjective community from the objects given in the immanence of subjectivity alone as units in the flow of inner time-consciousness. Hence, by distinguishing a positive from a negative criterion of transcendence, the egological reduction introduces "an abstract aspect within the residuum of the phenomenological reduction" (36). This abstract aspect includes in the residuum my actual perceptions along with the actual perceptions of possible Others in order to separate a positive sense of transcendence and so establishes "the strict separation between the sphere of ownness [*Eigenheit*] of subjective consciousness and the sphere of the intersubjectively given transcendent objects" (37).

In contrast, the primordial reduction is concerned with the actual givenness of Others. It brackets all contents that imply the givenness of Others, and its residuum is the primordial sphere of subjective consciousness. On this basis, Husserl attempts to show how an actual Other is given as a transcendent object.

A serious problem emerges at this point. On the one hand, the constitution of objects as transcendent in the positive sense entails the participation of Others. On the other hand, Others are constituted objects in the world. Transcendence presupposes intersubjectivity, and consciousness of the Other is consciousness of a transcendent object: "It is obvious that the whole enterprise becomes hopelessly circular if the givenness of the Other is treated on the same level as the givenness of real objects and the world" (Seebohm 1989, 375). Thus a strict separation between the experience of the transcendent world and the experience of the transcendental sphere of consciousness cannot be maintained. This means that the transcendence of the Other cannot be interpreted as the result of the constitutive activity of subjective consciousness and must be somehow incorporated into the transcendental sphere of consciousness. In this manner, it is shown that the ontic interpretation of the paradox of subjectivity overlooks the circularity insofar as it understands subjective consciousness as an absolute source in which Others are constituted.

In order to find a solution to the circularity, Seebohm considers that it is necessary "to draw some conclusions from Husserl's own descriptive investigations which he did not draw, or maybe even refused to draw" (375). An explication of transcendence must be developed, and Husserl has advanced materials with which problems raised by the transcendental reduction could be addressed. The application of the materials provided by Husserl with this purpose turns mainly on what Husserl has said about transcendence in immanence and on his analysis of the primordial sphere of consciousness.

Seebohm contends that intersubjectivity must be analyzed first of all in the light of a differentiation within the transcendental sphere of consciousness. Transcendence is given for consciousness at the genetically deep level of pure passivity before being the object of intentional acts. This is characteristic not only of the retentional continuum of the past but also of the experience of the Other. In the *Crisis* (1970, §54b), Husserl compares self-temporalization through de-presentation [*Ent-Gegenwärtigung*] with self-alienation [*Ent-Fremdung*], i.e., he shows a similarity between recollection and empathy as de-presentations of my primal presence [*Urpräsenz*] (1954, 189 [1970, 185]). The development of this transcendence of the Other in the immanence of the primordial sphere of consciousness must now be considered.

6.3 The Abstract Foundation of Consciousness

As the residuum of an abstractive reduction, the primordial sphere of consciousness is not "a concrete whole by itself," but rather "an abstract genetic foundation of and in individual consciousness in the world" (2015, 46). Following Husserl's analyses, a whole is a sum total of contents whose unity emerges out of a structured system of foundations between the contents or parts. Having this in mind, Seebohm stresses the following points in order to explain the genetic foundation of intersubjectivity within the structure of the primordial sphere (see Seebohm 2015, 37f., 45–48).[1]

(i) The primordial sphere of consciousness can be examined, in spite of not being a concrete whole by itself, as a lower-order dependent whole with its own dependent moments and foundations. It is a hyletic field whose contents are given through fusions and contrasts that emerge from passive syntheses and do not yet amount to objects. This primal sensibility is our first encounter with the brute reality of the pre-given lifeworld.

(ii) Dependent moments can be distinguished in the structure of inner time-consciousness as contents that change—the contents of the actual present, retention of past contents, and protentions of new contents—and in the structure of inner space as contents that move—the one Here and the manifold of Theres that surround it with their contents. Contents are given as changing in primordial time and moving in primordial space. There also emerges the contrast of the inner contents given in the inner self-givenness of the body in the Here and the outer contents given in the manifold of the There outside the body. One's own lived body is given from the inside in sensuous feelings and through its kinesthetic movements and from the outside in contents of primal sensibility.

(iii) Other living bodies are given from the outside in their movements, but their inner self-experience has to be appresented in an associative transfer based on similarities with one's own body. This appresentation enables us to understand the other living bodies in their life expressions by feeling through and with them. Thus one's own living body is experienced in a passive synthesis as a living body among other living bodies that are not given as correlates of intentional acts. Reciprocal action and reactions emerge by virtue of this first level of mutual understanding. Thus the Other as the other animate living body "is given in associative passive synthesis as a part of immediate brute reality" (Seebohm 2015, 47).

(iv) Objects shared in these interactions are given as transcendent in the immanence of the primordial sphere and within a shared intersubjective time and space. This distinction between primordial constitution and intersubjective constitution establishes a transcendence that goes beyond a unity in the immanent sphere of an individual consciousness. Hence transcendence emerges within the immanence of the primordial sphere as something immediately given for the subject. It is not the result of a constituting activity, and the givenness of intersubjectivity is also involved in its givenness in a reciprocal foundation within the realm of passive synthesis. This type of transcendence must also be distinguished from the transcendence given in subjective intentional experience. As the abstractive primordial reduction has bracketed the givenness of transcendent objects as correlates of active intentionality, this transcendence is

[1] These moments are described in the C-manuscripts and we shall turn to them later.

not the transcendence of an identical object given in different temporal perspectives under the presupposition of the possible actual presence of Others. No objective constitution is thus far involved.

(v) The ego emerges as a pole in active syntheses when configurations in the hyletic field awaken its attention and give rise to an active intentionality. Intentional acts are motivated by fusion and contrast phenomena in the hyletic field. Objects arise as correlates of this activity on the ground provided by the primal objects outlined in the realm of passive syntheses.

(vi) Understanding is the awareness of life expressions that indicate the inner life of Others and their involvement in intersubjective states of affairs. Its most basic form is animalic understanding in which the life expressions are grasped as actions that initiate reactions of one's own living body. Animalic understanding is founded on appresentation, includes expressions of animals and other human beings, and is limited to an immediate opposition of actions and reactions. This "bodily understanding" (Seebohm 2004, 103) is the genetic foundation for other types of understanding and is followed by elementary understanding, i.e., the understanding of how things are done in the interactions of practical life. We perceive roles and actions of Others, interactions with Others, and the materials, means, and goals included in this activity. All this adds up to a first-order elementary understanding of intentional as "an unreflective understanding within cultural contexts" (2004, 106) and is thematized by a second-order elementary understanding in which a comprehension of first-order understanding is undertaken.

The outcome of the foregoing analysis is that the structure of the primordial sphere with its subjective, intersubjective, and transcendent moments is pre-given for reflective analysis. The genetic analysis reveals that the emergence of the givenness of the Other and of transcendent objects "are correlated dependent parts of the givenness of the world," i.e., "are both together one-sidedly founded in the passive synthesis of contents of the hyletic field and in passive associative synthesis" (Seebohm 2015, 38). Intentional acts in direct intention give us objects and the world as a totality of beings, and the subject of these acts knows itself in its being in the world as incorporated in its living animated body with other animated and inanimated bodies: "objects and the world can be given in direct intention as transcendent correlates in the positive sense only because subjective consciousness is given to itself as a member of an intersubjective community in the world" (2015, 40).[2] We shall return to this later after dealing with another aspect of abstraction.

[2] Seebohm recalls that Alfred Schutz developed "an interpretation of the residuum that has the lifeworld as its residuum," so that phenomenological research is understood as "reflective analyses of the structures of the lifeworld" (Seebohm 2015, 85).

6.4 Methodological Abstraction in the Narrower Sense

Empirical sciences emerge in the pre-given lifeworld of elementary understanding. Having a one-sided generative foundation in the genetically previous lifeworld, they are part of a higher-order understanding that emerges as the theoretical understanding of what things are in the lifeworld. Higher-order understanding is manifested in cults, myths, prophetic revelation, poetry, painting, philosophy, and science. These cultural formations entail "the *creative* understanding of the context of the lifeworld as a whole and of partial aspects of the lifeworld" (Seebohm 2015, 76; see also 56). Thus natural sciences are part of first-order creative understanding of the lifeworld. They are founded in the encounter of elementary understanding with the natural environment, and challenge other "world views" by bracketing all other first-order interpretations of nature such as those offered by myth, poetry, etc.

The objects of an empirical science are intersubjectively accessible objects given in the lifeworld, and the ontological region that contains them is the result of bracketing certain aspects of the lifeworld. The "first axiom" of a phenomenological epistemology is that the establishment of the type of object pertaining to a specific science "presupposes a *methodological abstraction*, i.e., the determination of a limited ontological region of objects with a common formal and material categorical structure in the residuum of the abstraction…" (2015, 89). We can refer to a methodological abstraction in the narrower sense dealing with the methodology and objects of a specific empirical science in contrast to a methodological abstraction in the broad sense tied to phenomenological epistemology conceived as a general theory of knowledge. The narrower sense is Seebohm's main interest in his *History as a Science and the System of Sciences*.

The methodological abstraction described by Husserl in the *Crisis* must be understood in terms of two abstractions. A first abstraction is constitutive for the natural sciences in general. It determines the ontological region of the natural sciences in general by excluding all purposes of practical life, all ethical, aesthetic, and moral values of elementary understanding, and all other systems of higher-order understanding. Thus the natural scientist is a disinterested observer because all concerns except a theoretical interest have been set aside.

The second methodological abstraction determines the region of the hard sciences, i.e., sciences like physics and chemistry whose ontological categories can be explained in terms of mathematics. Only the primary qualities of objects are admitted because they can be legitimate objects for this kind of approach. This brackets the life sciences still included in the residuum of the first methodological abstraction. Darwin's theory of evolution, for example, remains outside this approach.

In a narrower use of the expression "reductive abstraction," Seebohm states that this second step is "a methodological abstraction that has in addition the character of a *reductive* abstraction" (2015, 170). With this he means that all phenomena relevant to other sciences are explained in terms of the categories of hard natural sciences. The point of making "the distinction between 'methodological abstraction' and 'abstractive *reduction*'" (2015, 380) is that the singled-out region of objects is raised to the condition of a basis for other regions of objects; for example, the objects of the life sciences and human sciences are epiphenomena that can be thoroughly explained away by theoretical entities of physics. That the second methodological abstraction also entails an abstractive reduction in this narrower sense is the claim of naturalistic physicalism: "What is given in the residuum of the first abstraction can be *reduced* to phenomena of the second abstraction. The second abstraction is, hence, an abstractive *reduction*" (2015, 222).

Seebohm contends that the abstraction that is constitutive of the natural sciences excludes certain essential structures of the lifeworld but does not set aside the lifeworld as such. There are aspects of elementary understanding in the lifeworld that play a role in the generative foundation for the basic concepts of the soft and hard sciences. Categories like matter, form, and causality can be mentioned here. For example, causality is involved in the understanding of changes in the regular course of nature as the result of divine processes and/or in the activities understanding the regular changes that follow the use of tools to reach a goal. Seebohm concludes: "Elementary understanding is, therefore, of basic significance for the explication of the foundations of the natural sciences in the lifeworld in general" (2015, 77). An analysis of the generative foundations for the emergence of the natural sciences must also include the interpretation of these concepts of elementary understanding in the level of higher-order understanding in pre-scientific philosophizing.

The first-order creative higher understanding, which is pre-given in a cultural tradition, calls for a secondary interpretation that amounts to, in Dilthey's characterization, to a "re-living" as "the *re-creative* understanding of the manifestations of first-order creative understanding in immediate and fixed life expressions, the *interpretation* of life expressions" (Seebohm 2015, 76; see also Seebohm 2004, 117). Second-order higher understanding of life expressions is the task of human sciences. Their theoretical goal is an interpretation of the meaning of the life expressions of contemporaries and predecessors. The methodological abstractions that determine a realm of observable objects as the region of human sciences excludes all aspects of the lifeworld given in first-order elementary and higher understanding and has as a residuum the region of second-order understanding of first-order elementary and higher understanding.

Life expressions of contemporaries given in the present are the objects of the systematic human sciences (psychology, social sciences), and fixed life expressions of predecessors are the objects of the historical human sciences (philology, history). An explanation of what life expressions meant in their past historical context requires a further methodological abstraction that brackets the original link in which interpretation emerges out of the need to restore communication and interaction in the lifeworld. All applications and rejections in the present temporal context of the interpreter are excluded (Seebohm 2015, 94). This methodological abstraction is an analogue to the abstraction that renders possible a disinterested observer in the field of the natural sciences.

History as a human science is guided by these methodological abstractions but transcends them because its reconstructions include dimensions of the lifeworld that belong to the natural environment. This turns it into a potential mediator between the natural sciences as sciences of explanation and the human sciences as sciences of understanding. This is a central thesis in Seebohm's *History as a Science and the System of Sciences*, but we cannot deal with it here.

6.5 Epistemic and Ontic Interpretations

Seebohm attempts "a critical explication of the structure and the problems of the paradox of subjectivity" (2015, 34). The paradox presupposes the ontic interpretation in which subjective consciousness is considered absolute and necessary in contrast to the contingent being of the world. The solution given by the epistemic interpretation presupposes the analysis of how oblique intention rests on direct intention and the explication of the emergence of transcendence in the positive sense within the primordial sphere. The following points are highlighted in Seebohm's argument (Seebohm 1989, 376f, 2015, 389f).

(i) Reflection has a one-sided genetic foundation in direct intention. The subject is not only given to itself in oblique intention, but is also, and primarily, given to itself in direct intention as a member of the intersubjective community. The subject of the intentional act in oblique intention is the same as the subject of the intentional act in direct intention.

(ii) An actual Now is the foundation of intentional acts in oblique intention, and the intentional act it reflects on is given with its object in the horizon of the past. Reflection can be iterated in a sequence in which the actual Now maintains its priority and the subject of the successive acts of reflection its identity. In the iteration, the intentional acts in oblique intention are always in the past horizon of the last reflection.

(iii) The actual Now of reflection is given in the future of a past actual Now as a not yet given Now. Seebohm describes it as "transfinite" in the sense that it surpasses the Now of intentional acts in the past horizon. It is given as "transfinite," but not as "absolute." If we are to speak of an "absolute ego," we must go beyond the limits of the epistemic attitude.

(iv) Analysis of the temporal structure of the relation between intentional acts in direct intention and intentional acts in oblique intention is the presupposition for a solution of the paradox of subjectivity. The paradox vanishes in the epistemic attitude if we focus on "what" is given in direct and in oblique intention to a subject that is not construed as an absolute ego in contrast to the contingent beings of the world. In direct intention, the subject is given to itself in presumptive evidence as incorporated "in" an animated living body that lives in the world together with other animated and inanimated bodies. In oblique intention, which has to presuppose this, the subject is given to itself in apodictic evidence as the subject of the givenness of the world. Through reflection, it is able to thematize and clarify all activities of consciousness. This gives the reflective attitude an epistemic priority that contrasts with the genetic priority of the natural attitude in which phenomenological reflection is one-sidedly founded.

To sum up: the subject knows itself at the beginning of this epistemic analysis as part of the world and at the end gains no access to an absolute ego: "A *reflective* phenomenological analysis is as such *not* able to analyze the 'awareness' of its own actual Now as opening an access to the absolute transcendental ego… . The experiences of a phenomenology of a higher order that is required for a speculative metaphysical experience is irrelevant for epistemological phenomenological reflections" (Seebohm 2015, 389f).

That the ontic interpretation is considered irrelevant for a phenomenological epistemology does not mean that it is meaningless (Seebohm 2015, 387, 390). Seebohm contends that the ontic interpretation is a method for the immediate and speculative experience of an absolute ego beyond the ego given in reflection as a pole for the manifold acts of consciousness: "Such an ABSOLUTE EGO is the ultimate ground, and as an absolute *entity*, it posits the emergence of an ego as a pole together with its correlate, the not yet understood brute reality of the hyletic field, and with it the world in the actual standing Now of inner temporality" (2015, 388). Seebohm contends that, although there is textual evidence that Husserl was inclined to a quasi-Fichtean metaphysical idealism, other interpretations of the primal ego are possible.

With the foregoing enunciation of the different moments of the epistemic attitude in mind, let us now turn back to Husserl. Some aspects of his discussion of the sphere of primordiality will allow us to deal with the ultimate ground in a more inclusive manner. What is attempted is a further application of the materials offered specially by the C-manuscripts to the examination of the ultimate grounding of transcendental phenomenology.[3] This application takes us beyond Seebohm's elucidation of the primordial sphere.

Two sides must be distinguished when an ultimate ground is referred to. In a passage of the C-manuscripts, Husserl states: "Regarding primality [*Urtümlichkeit*] it is natural to distinguish the primality of myself inquiring back from the constituted world—of myself, the mature I reflecting on myself—and the primality that emerges out of the further regressive inquiry and is reconstructed through uncovering the 'beginning' of the constitutive genesis" (2006, 279). Latent in this indication of a two-fold primality is the distinction between the primal ego [*Ur-Ich*] and the pre-ego [*Vor-Ich*], even if they are linked as manifestations of the same ego and Husserl's use of these terms is at times ambiguous. The primal ego functions permanently in the living present, whereas the pre-ego has to be reconstructed by inquiring back

[3] It is worth remembering that, in the *Cartesian Meditations*, Husserl lays stress on the fact that generative problems "belong to a higher dimension and presuppose such a tremendous labor of explication pertaining to the lower spheres that it will be a long time before they can become problems to work on" (1950, 169 [1960, 142]).

into the past in order to show the link between subjectivity, instinct, and teleological drives (Taguchi 2006, 119). According to Husserl, the mature ego reflecting on itself in the transcendental attitude not only gains access to the pre-ego but also discloses the primal ego as the ultimate condition of possibility of all our experiences and activities. The primal ego is presupposed in all passive and active syntheses as the ultimate core that accomplishes every synthesis. It is prior to all acts and sense data, to all awareness of successive events, to Others, and to the ego itself as part of an intersubjective community. Nevertheless, as we shall see, it opens out to intersubjectivity prior to any actual experience of the Other.

Seebohm presents the difference between the epistemic and the ontic attitude as "a difference in the apprehension of inner time consciousness" (Seebohm 2015, 389). It is the difference between focusing on a "transfinite Now," i.e., a Now that is anticipated but not yet given, and a permanent Now that as a *nunc stans* separates itself from all passing finite Nows (see Husserl 2006, 8; Seebohm 1989, 370). The problem of the permanent Now is overlooked, but the matter is not quite so simple. In Husserl's view, this permanence is demanded in order to take account of the self-manifesting synthesizing function of the primal ego in time syntheses and must be completed with an analysis of the depth dimension of the ego explaining how the protentional drive rests on the instinctive impulses of the pre-ego. Thus the reference to an ultimate foundation requires certain qualifications. If the transcendental ego as an abstract pole in syntheses is "an indicator of the transcendental ego understood as the self-manifesting activity of the *Ur-Ich*, the 'primal ego,'" (Seebohm 2015, 386) the absolute ego in turn is not "independent and unconditioned in every respect," because, as an abstract moment, it is also the indicator of a concrete whole that includes the pre-ego.

Seebohm's account conflates two coexistent but different aspects of the ego. In the following sections an attempt is made to show why both aspects of primality must be taken into account. In the next section we will see that the genetic process *evolves* from the pre-ego, through the awakened ego as a pole of acts with its primal ego as their ultimate source, up to a kind of reflection that, emerging from the primal ego, *involves* the whole process in the sense of disclosing its transcendental significance. In the final section it will be shown that, along with the givenness of subjective consciousness to itself as a member of the intersubjective community, a certain kind of givenness of the Other can also be unfolded both in the initial level of the pre-ego and in the mature level of the primal ego. This means that there is more to Seebohm's elucidation of the primordial sphere.

6.6 Ontological Themes

Husserl's retrospective inquiry appeals to a "method of reconstruction" that allows significant complements to Seebohm's analyses of the primordial sphere as a hyletic field with the examination of a "concrete animalic ontogenesis and phylogenesis"

(2006, 170, 352; see also 1973, 595) and the explication of the instinctive background of the pre-ego. A "transcendental life-stream" is originally guided by instincts with drives and strivings whose satisfaction outlines a teleology. At the initial stage of human experience, two sides can be abstractly separated within the undifferentiated whole of this life-stream. On the one side, the pre-ego functions as the focal center that individualizes in a genetic process. On the other hand, "the universe of previous being" [*das Universum des Vor-Seienden*] is also characterized as a "hyletic quasi-world, alien to the ego" or a "previous world" [*Vor-Welt*] (see 2006, 22, 187, 199, 350). Husserl goes back to a pre-perceptual and pre-objective level of experience in which there is no discrimination within a mass in which a primal hyle on the non-egoical side is fused with a primal feeling and primal kinesthesis on the side of the pre-ego. Out of this life-stream, distinct objects will eventually emerge: "In the genetic retrospective inquiry we construct as a beginning the previous field and pre-ego [*Vorfeld und Vor-Ich*], still without world, which is already a center, but still not a 'person', let alone a person in the habitual sense of human person" (2006, 352; see also 2006, 198f., 335). As described above in Sect. 6.3, an active ego is awakened by hyletic configurations and begins the constitutive process by bestowing meaning on such hyletic fusions and contrasts, which are "affecting onta," in "the first level of ontification or objectivation" (2006, 198, 252). From then on, through a plurality of levels with their relative worlds, "the world in a proper sense" [*die eigentliche Welt*] is constituted (2006, 350).

The ego ignores itself as transcendental because it lives "in the mode of naïve hiddenness" (1954, 214 [1970, 210]). Although still hidden to the ego, the constitutive development is teleologically directed to a stage in which it becomes aware of itself and the true nature of consciousness and its constituting processes through a radical reflection. Husserl speaks of a "potentiality to break through from my being valid as a human being to transcendentality" (1973, 486f). The apodicticity of the evidence in the givenness of consciousness to itself has an ontological presupposition because a previous process is necessary. Once the evidence is attained, it enables consciousness to bracket, as Seebohm holds, all the claims for objectivity that are not given in apodictic evidence and to elucidate them from its own standpoint. Furthermore, Husserl's method of reconstruction allows an unlimited inquiry back and shows that the genetic priority assigned by Seebohm to the natural attitude can be elucidated from a transcendental point of view by appealing to a constructive phenomenology. This is required when the given and its horizons provide a motive for further investigation (see Fink 1988, 61–74 [1995, 54–66]). This retrospective inquiry leads us back in the paths of the above-mentioned phylogenesis and ontogenesis. Consequently, that "a universal phenomenological reflection is universally one-sidedly founded in the natural attitude" (Seebohm 1989, 377). does not exclude the possibility of a transcendental approach to what is given along with an appraisal of its ontological precedence.

The development from the pre-ego in the instinctive level up to the stage in which the primal ego is disclosed in a radical reflection is not devoid of ontological significance and suggests steering a third way between the seemingly extreme poles of epistemic and ontic interpretation.[4] The ontological significance is involved not only in the pre-world to which we are led back as a presupposition of the development that has to be reconstructed but also in the process of ontification achieved by meaning-bestowing acts that have to be unfolded. Two aspects of meaning-bestowal deserve to be brought out with regard to Seebohm's interpretation. They concern the relevance of ontification in direct intention and in the founding level of empirical science. Thus, in contrast to Seebohm's view (1989, 364, 370), the "ontological question of being" is not "absurd" and cannot be set aside even if the main interest lies in an epistemology of empirical sciences.

First of all, the transition from concealment of this function in the natural attitude to its disclosure in transcendental reflection is an important point that should not be neglected in the analysis of the relation between direct and oblique intention. For both kinds of intention can be performed in the natural attitude, in which meaning-bestowing acts remain concealed, and in the transcendental attitude, in which they are made known. Seebohm's association of the natural attitude with direct intention and phenomenological attitude with oblique intention does not allow for the fact that transcendental consciousness, blind to its transcendental dimension, turns to the world in direct intention and for the further fact that one can also enter into the oblique intention in the natural attitude. The transcendental character of the ego is

[4] As Seebohm offers "a systematic survey of the epistemic and ontic interpretations in the development of the phenomenological movement" (2015, 43), it is helpful to add a scheme outlined by Elisabeth Ströker on the relationship between consciousness and the living body (see 1977, 168–171). From an epistemic point of view, consciousness has a priority because any attempt to state reasons for its derivation presupposes consciousness itself and hence entails a fallacy. From an ontological viewpoint, the relationship is complex. On the one hand, there is a two-fold conditioning. An ontological conditioning of consciousness by the living body means that consciousness is unthinkable without a living body because it depends on the factualness of a living body even if it is not contingent upon the individual factualness of a particular living body. And an ontological conditioning of the living body by consciousness means that a living body is always dependent on consciousness because it is teleologically oriented toward awareness of itself through reflection. On the other hand, there is also a two-fold non-derivation. The living body cannot be derived from consciousness because transcendental constitution is confronted with a residuum that cannot be resolved analytically into its temporal, associative, and kinesthetic syntheses. There is a remainder, a factum or brute reality, that cannot be dealt with conclusively in terms of an analysis because it cannot be circumvented and so is not at our disposal (see Ströker 1987, 73 f.). And consciousness cannot be inversely derived because the living body only shows "how" it is possible but cannot account for "what" consciousness actually is. The "what" can only be revealed by consciousness itself when it becomes aware of itself in reflection. Although the scheme goes against the statement that "the ontological is not a theme for phenomenology" (Seebohm 1989, 377), some coincidences can be shown. Along with a convergence of the epistemic priority with Seebohm's claim on "the epistemic priority of phenomenological investigations over all other types of investigation" (366), other parallels can be drawn. The ontological conditioning of consciousness and the non-derivation of the living body can be associated with Seebohm's assertion that "the natural attitude is genetically one-sidedly founding of the transcendental attitude and has genetical priority" (Seebohm 376f). The ontological conditioning of the living body by consciousness can be linked to the striving of consciousness for evidence and the iteration of oblique intentions. And the non-derivation of consciousness from the living body amounts to Seebohm's rejection of the abstractive reduction of all the contents of lived experience in the lifeworld to the theoretic objects of the hard sciences.

independent from the distinction between direct intention and oblique intention because it belongs both to the transcendental spectator that discloses the origin of world-constitution and to the subject that, although blind to constitutive performances, intends the world in meaning-bestowing acts.[5] Husserl holds that the phenomenological reduction enables us to recognize the "function of forming ontic meaning" [*Seinssinn-bildende Funktion*] (1954, 172 [1970, 168]). Since reflection presupposes meaning-bestowing or ontifying acts in direct intention, no epistemology in the broad sense would be possible without them.

In addition, an analysis of the meaning-bestowing function in the founding levels is of primary importance. Meaning-bestowing is carried on before the emergence of science and deserves particular attention as a process of ontification, especially if elementary understanding in the lifeworld plays a role in the genesis of the basic concepts of science. It is also a part of the generative foundation and as such requires a specific analysis that weakens the force of an epistemic approach exclusive of ontological implications. In Husserl's view, transcendentalism means that "the ontic meaning [*Seinssinn*] of the pre-given lifeworld ... is the achievement of experiencing pre-scientific life" and that, in this life, "the meaning and ontic validity [*Seinsgeltung*] of the world are built up... ." (1954, 70 [1970, 69]). Again, without this ontic meaning and validity pertaining to the meaning-bestowing or ontifying acts of elementary understanding there would be no founding level and motivation for the higher-order understanding that characterizes empirical sciences.

6.7 The Pre-ego, the Primal Ego, and the Internality of Subjects

The structure of the primordial sphere involves, as Seebohm stresses, subjective, intersubjective, and transcendent moments that make up a whole with their reciprocal foundation, but there is still a privileged moment in this domain. The primordial reduction is defined by Husserl as "the unbuilding of appresentations that cannot be turned into my genuine presentations" (1973, 125). Here he alludes to appresentations whose perceptive fulfillment does not depend on the movements of my living body and the perceptions that are motivated in this way. When these appresentations are bracketed, a domain of my exclusive experience is delimited and contrasted with the experience of Others. It is by reason of this nucleus of original presence that the mutual foundation of subjective, intersubjective, and transcendent moments in the primordial sphere is unbalanced to the benefit of subjectivity. Starting from this prominent moment we gain access to Others through apperception and empathy.

[5] Fink notes that "the onlooker is only the *functional exponent* of transcendentally constituting life, an exponent that of course does not itself in turn perform a constituting action but precisely through its transcendental differentness makes self-consciousness (becoming-for-oneself) possible for constituting subjectivity" (1988, 44 [1995, 40]). Husserl only introduces in this text minor alterations that do not affect the main idea.

However, the access is conditioned by an internality or intermingling [*Ineinander*] of subjects due both to intersubjective instincts and to a particular previous appresentation linked to the primal ego.

As a pre-ego, each monad is situated in "the totality of monads in an originary instinctive communication" (1973, 609) by virtue of "an instinct that leads to the cosubjects and subjects taken universally" (2006, 338). One can go beyond Seebohm's analyses by including in the brute reality that can be found in the primordial sphere an instinctive reference to the copresence of Others. Husserl alludes to the development of each individual ego out of transcendental instincts that entail relatedness to Others and are the teleological ground for an under-egoical temporalization in an intersubjective previous time [*Vorzeit*]. He refers to a universal impulse that composes each present and makes the flow of time move along. The various monadic life-streams are founded in "a nexus of intersubjective intentionality that, taking place certainly in the individual subjects, is at the same time a unitary intentionality, …" (Husserl 2013, 440; see also Husserl 1973, 595ff.). Within the "coexistence and competition of particular instincts in their mode of outcome" (2006, 258), a "transcendental instinct" is also an "instinctive impulse of objectivation" (2006, 258, 260, 331). Thus, since transcendence and intersubjectivity are both sustained on the same basis, they are not in a circular relation of presupposition in which the constitution of one depends on the constitution of the other.

The primal ego also carries with it an original relation to the Other. Husserl writes: "The Other is copresent in me… . I, as absolute, as a living and flowing existent and concrete present have her/his present as copresent, as appresentatively-manifesting-itself-in-me (*als Appresentativ-sich-als-es-selbst-bekundend-in-Mir*), … . I carry all Others in me, themselves as appresented and to be appresented and as carrying myself in themselves" (2006, 56; see 56ff.). The "coexistence [*Mitsein*] of Others" means that "this copresent of Others is founding for a worldly present, … ." (2006, 57). Here we must recall that, as mentioned previously, Husserl compares self-alienation in the experience of Others with depresentation in time-consciousness. The analogy can be unfolded as follows.

Having in mind a retention that is not directed to past phases of our own acts or sense, we can turn our attention to an appresentation that is not directed mediately to the Other through the experience of a living body. In his earlier lectures on time, Husserl made a distinction between the retention of the phases of experiences or inner objects through transverse intentionality and the retention of the phases of an absolute flow through horizontal intentionality. By virtue of this double intentionality, not only do we retain past phases of our experiences, or inner objects of our stream of consciousness, but also the elapsed phases of our consciousness of these phases (see Husserl 1966, 80–83, 116–118 [1991, 84–88, 120–122]). In the case of appresentation, its scope is broadened out to include an originary immediate copresence of Others. This would mean an intentional reference to the Other without the underlying mediation of the perception of an alien living body. In an analogous "double intentionality" appresentation in the usual sense would have the ground of its possibility in a primal appresentation that accounts for unity among streams of consciousness in the same way as the horizontal basic intentionality of retention accounts for unity within the absolute flow. In other words, Husserl is concerned here with an appresentation that cannot be bound up with a presentation in the sphere of outer experience. This implies that it does not point back to an analogizing apprehension of another living body but rather to a network of implications pertaining to the primal ego in the living present. Its motivational basis is not the similarity connecting the body over there with my living body but a similarity connecting the implication of a being otherwise of the ego as the same ego with the implication of being otherwise of the ego as another ego.

We have therefore the two following full sequences that explain the parallel between depresentation in time and self-alienation. In my stream of consciousness I have the sequence "retention of an elapsed phase of the absolute flow / retention of an elapsed phase of an experience (sense data, acts)/recollection of a past act with its ego (my being otherwise). The series has a counterpart in the sequence "appresentation implicated within my absolute flow/appresentation motivated by the perception of another living body/empathy of an act of the Other (a being other than me)." This provides an answer to Husserl's rhetorical question: "Does there belong to every primal continuation (retention) also a primal empathy [*Ureinfühlung*], or, should I say instead of an empathy that is explicating, a primal intentionality of the manifestation of continuity with Others, which, as the temporal fusion, is appresentative [*ad-präsentierende*] in a mediate manner and in continuous mediation?" (Husserl 2006, 437).

The outcome of this analysis is that one cannot ascribe to Husserl a reading of the ontic interpretation of the reduction in the light of a strict separation between the experience of the transcendent world, which entails the participation of Others, and the experience of one's own transcendental primordial sphere, which excludes Others. First, the transcendental subject is referred to in an instinctive communication at the level of the pre-ego. Second, the unique and indeclinable primal ego breaks through to an intersubjective dimension by way of a "horizontal appresentation." Functioning as a primal ego in its unfolding and disclosure in reflection, the ego maintains in a depth dimension the intersubjective reference it always carries as a pre-ego. Only an abstraction can separate the pre-ego from the conditioning character of intersubjective instincts and instinctive kinestheses—the living body—and the primal ego from the appresentative implication of the copresence of Others. Instinctive communication on the one hand and primal empathy—or primal appresentation—entailing "continuity with Others" on the other hand are an inseparable part of the ego.[6] Thus, there is no circularity in the constitution of an objective world and Seebohm's one-sided emphasis on the epistemic question about modes of

[6] The implication of Others should not lead one to overlook the fact that the ego maintains a primordial sphere with an exclusive right of ownership, starting from which, and guided by this communication and continuity, he gains access to Others through second-order appresentation and empathy. This goes hand in hand with, and has its ultimate root in, the uniqueness and indeclinability of the ego in the sense that, "starting from itself and in itself" [*von sich aus und in sich*], the ego "constitutes transcendental intersubjectivity to which it then adds itself as a merely privileged member, namely, as 'I' among the transcendental others" (1954, 188 [1966, 185]; see 1973, 586; 2006, 386).

givenness with the omission of ontological issues cannot be accepted without qualification. We have seen that the epistemic interpretation and the ontic interpretation leave room for one another and even condition each other. Ontological issues have emerged, and receive a transcendental reconstructive or disclosing analysis, with regard to phylogenesis and ontogenesis in monadic development, the function of forming ontic meaning in human history, and coexistence of Others.

References

Fink, E. 1988. *VI. Cartesianische Meditation. Teil I*, Husserliana-Dokumente II-1, ed. H. Ebeling, J. Holl, and G. van Kerckhoven. Dordrecht: Kluwer.

———. 1995. *Sixth Cartesian meditation.* Trans. R. Bruzina. Bloomington: Indiana University Press.

Husserl, E. 1950. *Cartesianische Meditationen und Pariser Vorträge*, Husserliana I, ed. S. Strasser. The Hague: Martinus Nijhoff.

Husserl, Edmund. 1954. *Die Krisis der europäischen Wissenschaften*, Husserliana VI, ed. W. Biemel. The Hague: Martinus Nijhoff.

Husserl, E. 1960. *Cartesian meditations*. Trans. D. Cairns. The Hague: Martinus Nijhoff.

———. 1966. *Zur phänomenologie des inneren Zeitbewusstseins (1893–1917)*, Husserliana X, ed. R. Boehm. The Hague: Martinus Nijhoff.

———. 1970. *The crisis of European sciences and transcendental phenomenology*. Trans. D. Carr. Evanston: Northwestern University Press.

———. 1973 *Zur Phänomenologie der Intersubjektivität.Texte aus dem Nachlass, Dritter Teil (1929–1935)*, Husserliana XV, ed. Iso Kern. The Hague: Martinus Nijhoff.

———. 1991. *On the phenomenology of the consciousness of internal time (1893–1917)*. Trans. J. B. Brough. Dordrecht: Kluwer.

———. 2006. *Späte Texte über die Zeitkonstitution (1929–1934). Die C-Manuskripte*, Husserliana-Materialen VIII, ed. D. Lohmar. Dordrecht: Springer.

———. 2013. *Grenzprobleme der Phänomenologie. Analysen des Unbewusstseins und der Instinkte. Metaphysik. Späte Ethik. Texte aus dem Nachlass (1908–1937).*, Husserliana XLII. Ed. R. Sowa and T. Vongehr. Dordrecht: Springer.

Seebohm, T. 1989. Transcendental phenomenology. In *Husserl's phenomenology: A textbook*, ed. J.N. Mohanty and W.R. McKenna, 345–385. Washington, DC: University Press of America.

Seebohm, T.M. 2004. *Hermeneutics: Method and methodology*. Dordrecht: Kluwer.

———. 2015. *History as a science and the system of sciences: Phenomenological investigations*. Dordrecht: Springer.

Ströker, E. 1977. *Philosophische Untersuchungen zum Raum*. Frankfurt: Vittorio Klostermann.

———. 1987. *Phänomenologische Forschungen*. Frankfurt: Vittorio Klostermann.

Taguchi, S. 2006. *Das Problem des 'Ur-Ich' bei Edmund Husserl*. Dordrecht: Springer.

Roberto J. Walton (Ph.D., University of Buenos Aires) is Professor at the University of Buenos Aires and member of the National Council for Scientific Research. He is the Director of the Center for Philosophical Studies at the National Academy of Sciences in Buenos Aires. He is the author of *Husserl: Mundo, conciencia y temporalidad* (Buenos Aires: Almagesto, 1993), *El fenómeno y sus configuraciones* (Buenos Aires: Almagesto, 1993), *Introducción al pensar fenomenológico* (with A. Ales Bello, Buenos Aires: Biblos, 2013), *Intencionalidad y horizonticidad* (Bogotá, Aula de Humanidades, 2015), and *Horizonticidad e historicidad* (Bogotá: Aula de Humanidades, forthcoming).

Chapter 7
The First Specific Abstractive Reduction in Seebohm's Theory of Science

Lester Embree

Abstract The *Crisis of European Sciences and Transcendental Phenomenology* and earlier writings mention *prima facie* only one abstractive reduction for the natural sciences in general. Considered more closely, however, two abstractions can be distinguished. The first reduction abstracts from the purposes and values of everyday practical life, and in general from all contents that are given in lived experience and reflections on lived experience. The second abstractive reduction abstracts from all secondary qualities of observable objects that are given in the residuum of the first abstraction, reducing them to primary qualities. The first abstraction determines the ontological region of the natural sciences in general, including the life sciences. The second abstraction determines, within this broader region, the empirical basis and the methodology of the hard sciences, physics, and chemistry insofar as it can be reduced to physics.

7.1 Introduction

There is considerable technical terminology in *History as a Science and the System of the Sciences: Phenomenological Investigations* (Seebohm 2015, 6) and some deserves comment. For example, "direct" and "oblique" are preferred over "straightforward" and "reflective" (2015, 15). Much clarification is called for. I begin with my title and will include much quotation in footnotes as well as in the text of this study.

Alfred Schutz actually uses *Wissenschafislehre* just once, but, following my teacher Dorion Cairns, I have rendered it repeatedly in my studies of Schutz as "theory of science" or "science theory" (cf. Embree 2015) chiefly because some scientists also reflect on the distinctive methods, basic *the*orems, and disciplinary definitions of their particular disciplines, and it is odd to say they then do philosophy of science when they are seldom well-prepared in that subdiscipline of

L. Embree (✉)
Florida Atlantic University, Boca Raton, FL, USA
e-mail: embree@fau.edu

© Springer Nature Switzerland AG 2020

T. Nenon (ed.), *Thomas Seebohm on the Foundations of the Sciences*, Contributions to Phenomenology 105,
https://doi.org/10.1007/978-3-030-23661-8_7

philosophy. While Seebohm speculates about Schutz's preference of *Wissenschaftslehre*[1] (which Husserl used) over *Wissenschaftstheorie*, he accepts this English translation[2] and only uses "philosophy of science" three times.

Next, I would prefer to use "*epoché*" because, on the basis of oral tradition from Dorion Cairns in a conversation with Eugen Fink (unfortunately not documented or applied by Cairns), I can then restrict "reduction" to changes in attitudes and "purification" to changes in intentional objects. But Seebohm prefers "abstractive reduction" and I usually follow that.

While the transcendental phenomenological reduction can be epistemically considered *generic* (Seebohm 2015, 94, cf. 86), and thus above any specific abstractive reductions, botany and zoology and astronomy and chemistry, for example, can be considered particulars under the biological and physical species. There are of course many particular sciences. This taxonomy is implicit but not deeply so.

What I seek here to show is that there are actually three rather than two *specific* abstractive reductions. Since Seebohm regularly qualifies what I consider the second and third as the "first" and "second" species and a new numbering would be confusing, I will regularly just name these abstractive reductions as "the cultural-scientific reduction," "the biological-scientific reduction," and "the physical-scientific reduction."

The epigraph above is thus inadequate or at least misleading because what one can first get starting from the lifeworld is the field of what, being less of a traditionalist than Seebohm,[3] I prefer to call the "cultural sciences" chiefly because some non-human species, e.g., chimpanzees, as Scheler, Gurwitsch, and MerleauPonty knew, can be said to have non-verbal culture and thus that interesting comparative

[1] "The difficulty for Schutz in the cultural context of the United States, which was governed at this time by analytic positivism, was in effect to explain the difference between a *Wissenschaftslehre* and a *Wissenschaftstheorie,* i.e., a theory of science. A theory of science is an *epistemology* of positive sciences in general" (Seebohm 2015, 307, cf. 318).

[2] "The present investigation is interested in 'epistemology' in the narrower sense. It is a phenomenology of *scientific* knowledge of the *empirical* sciences and their methodology as a part of a theory of scientific knowledge in general, i.e., of a *Wissenschaftslehre,* a theory of science, and a theory of scientific knowledge is in tum a partial theory of knowledge in general, including the knowledge implied in the elementary understanding of practical knowledge. In contrast, a phenomenological epistemology in the broad sense is interested in *all* subjective intentional acts and intersubjective activities because all of them imply 'knowledge,' understanding, of their intentional objects as their correlates. Thus, it includes the phenomenological reflections on intentional acts and syntheses in the practical activities of elementary understanding and in systems of first-order higher understanding such as religion, poetry, arts, etc., that refer to practical actions and interactions" (Seebohm 2015, 88, cf. 307f).

[3] "Schutz, following Rickert, sometimes preferred *Kulturwissenschaften,* cultural sciences. From the epistemological point of view, this term has the advantage of avoiding the metaphysical connotations of the te1m *'Geist'* but also of connotations of 'human' as a predicate used in te1ms for branches of the natural sciences or the technological application of natural sciences, e.g., in medical technologies. The terminology of the following investigations will use 'human sciences' because it is not advisable to 'deconstruct' systems of already established terminological traditions in the sciences" (Seebohm 2015, 71 n.15, cf. 384).

investigations with humans are possible. Seebohm has statements about "animalic understanding" using dogs as examples. Again, his preference of "human sciences" will usually be followed to reduce confusion.

Seebohm also uses the expression, "methodological abstraction" and probably does so twice as many times as "abstractive reduction." One might suppose them synonyms, but he is explicit that there is a distinction between these two abstractions (Seebohm 2015, 380). While I could not find a concise definition for "abstractive reduction," a statement about the "methodological abstraction"[4] includes "ontological" and makes this one relevant:

Two interpretations are possible here, the *epistemic* interpretation and the *ontic* interpretation. The epistemic interpretation is sufficient for the purposes of a phenomenological epistemology. Of interest for the epistemic interpretation are only the scope, the limits, and the quality of the evidences in which the objective correlates of synthetic intentional activities are given in direct and oblique intention. The ontic interpretation is grafted upon the epistemic interpretation. Epistemic categories are understood as indices of a metaphysical ontology speaking about necessary being, contingent being, absolute being, etc. (Seebohm 2015, 38).

That Seebohm did not exclude ontology is clear, e.g.: "A rough survey of the historical development of ontological reflections on the life sciences can serve as a preparation for the phenomenological analysis of these categories" (Seebohm 2015, 240).

Phenomenologically, I hold the *epoché* for the human sciences consists in the adoption by the scientist of the theoretical attitude and getting a field purified of her own everyday purposes and values. Does Seebohm actually recognize this? To show that he does most of this study is not a phenomenological but philological task. I begin with the lifeworld.

7.2 The World of Everyday Life

Where must we start from in order to do science? It is obvious but often overlooked that we can only start from where we are when not doing science or even philosophy. Where we originally are is living our everyday lives in the socio-historical world of culture, also known as "the lifeworld" or, better, "*a* lifeworld," for there are many, although all lifeworlds exemplify a generic *eidos* that the phrase "socio-historical world of culture" at least begins to describe ("cultural lifeworld" is then

[4] "A methodology of a science has to determine the type of admitted theoretical constructs guiding the search for promising hypotheses. It determines, secondly, the basis, the type of objects that can count as objects of the science. This determination presupposes a *methodological abstraction,* i.e., the determination of a limited ontological region of objects with a common formal and material categorial structure in the residuum of the abstraction, with all other types of objects always bracketed. The universal realm in which such abstractions are possible is the lifeworld" (Seebohm 2015, 89).

nicely redundant). In the above epigraph, there is reference to "the purposes and values of everyday practical life," which Seebohm attributes to Husserl (Seebohm 2015, 281). It is traditional to contrast practical and theoretical life, but since there is not only the pursuing of purposes but also enjoyment and suffering in everyday life, Husserl's inclusion of values is apt.

Seebohm calls our own particular lifeworld a "lifeworld with sciences" (2015 n 60, 77, 107, 154, 232, 236). He also holds that such a lifeworld is a precondition for the human sciences, economics in particular (332), and claims it is "necessarily a lifeworld with scientific technologies" (182), and that "Natural science has … a value because for elemental understanding it is a tool, an instrument for the realization of practical purposes and goals" (236). This is an advance and deserves development on another occasion. He also recognizes foreign lifeworlds and lifeworlds that are "pre-scientific" in the signification of being historically prior to our lifeworld.

Seebohm is especially interested in the historical human sciences and that interest centers not only on one type of higher understanding but also on what he calls "fixed life expressions," i.e., texts and traces. But for the question of what the world is in our everyday life, i.e., apart from the adoption of any contemplative, theoretical, or disinterested attitude toward it, there is need to recognize lower types of understanding and another type of expression.

The section entitled "A Typology of Understanding" in the new book is short, just over three pages, but refers to the earlier book, *Hermeneutics: Method and Methodology* which has a more extensive presentation of the "immediate" and "fixed life expressions" and the "animal," "elementary," and "higher" understanding.[5]

Texts and traces are crucial for hermeneutics, but immediate life expressions also matter. "Immediate life expression" occurs in the earlier book: "Present life expressions in general include fixed and unfixed life expressions. Both are in part linguistic and non-linguistic, like a command or a threat with a weapon, a promise, a dance" (Seebohm 2004, 83). Presumably the "unfixed" are the same as the "immediate." "Both gestures, which are pure bodily life expressions, and other immediate life expressions will be possible referents of discourse if they are relevant for the interactions in elementary understanding, e.g., bodily life expressions indicating willingness or unwillingness to cooperate or then ability/inability to perform certain activities because of pain, hunger, signs of rage indicating disruptive behavior, or the like" (111, cf. 114).

According to the new book, "Immediate life expressions have their genetic foundation in animalic life expressions" (2015, 49) and there are several statements about "animal understanding," for example:

> This rudimentary understanding can be called bodily understanding or animal understanding. It can be called 'animal' because in the everyday experience in the lifeworld … we

[5]Wanting a more technical sounding expression, Seebohm found "animalic" in a dictionary and introduced it later (Seebohm 2015, 7 n. 5).

assume that animals are able to understand us and each other in basically the same way as we understand them. In the shared time/space of the living present we have, however, shared objects that each of us senses from different perspectives, perspectives that change in accordance with mutually observable kinaestheses. They are on this level first-order, intersubjectively given real objects. Not only a stick in the shared time of a dog and his master playing in the garden, but a bone about which two dogs are fighting is already such a shared object. (2004, 103)[6]

"First-order animalic understanding" is further elaborated elsewhere in the new book outside the section on the typology of understanding, but with "secondorder animalic understanding" not mentioned (2015, 392)![7] The correlation on this lowest of levels is clear: "First-order animalic understanding by itself is originally given in communications via immediate bodily life expressions" (323).

Central, however, to Seebohm's account of the everyday lifeworld is "elementary understanding." The word occurs over 300 times in the new book, but most of what is relevant is said about it already in Sect. 3.2. While animalic understanding is Seebohm's addition, elementary understanding comes from Dilthey and animalic understanding is Seebohm's addition. It is "approximately coextensive" with "practical world" in Schutz, an expression Seebohm uses regularly. Both immediate and fixed life expressions are understood in it.

Elementary understanding is the understanding of actions, of interactions with others, of others as participants in interactions, of the material, the means, and the goals of actions and interactions. The raw materials offered by the *natural environment* in a lifeworld are understood "as good for something," as means and goals in the practical world. The means in actions and interactions are tools. The products of using tools are artifacts used for the fulfillment of the needs of everyday life. Tools of humans are seldom found in the natural environment as materials that are as such immediately useful for something; instead, tools are themselves artifacts produced by other tools. Actions are understood as actions in which "how to do something" with tools is understood and Others are understood as participants in interactions, i.e., they are understood only in their functions and roles in the web of elementary understanding.[8]

---

[6] "Animal interaction requires in addition certain messages between living beings. The messages guiding animal understanding are bodily life expressions. Some of them are not deliberately addressed to another currently living being, e.g., the smell and bodily behavior of a bitch in heat, while others are indeed a reaction to an encounter with another living being and are addressed to this other living being with the intention of ... causing a celiain reaction, e.g., the growling and snarling of a dog. The ability to interact with the life expressions addressed to it is different in different species [and] the ability of a bodily being to understand, i.e., to react properly in interactions, increases with the degree of similarity between the living bodies, and is of the highest degree within one's own species... . Sending and understanding bodily life expressions should not be called 'body language' but 'non-verbal bodily communication'" (2004, 104).

[7] "First-order understanding" occurs five times (2015, 298, 300, 323, 349, 406) but "second-order animalic" occurs nowhere explicitly. Perhaps it is implied in some places.

[8] "Tools" are defined broadly as "instruments that can be used to introduce changes in the course of events..." (Seebohm 2015, 274, cf. 270) and are related to techniques. Thus, there are scales (219), "the experience of weight is already an outstanding necessary part of the structure of the lived

The system of elementary understanding is a system of goal-directed interactions. Successful interactions presuppose immediate life expressions that can serve as signifiers for all elements of the system before and after the interactions happen in the actual present. Such a sign system is a linguistic system in the broadest sense. Human discourse is originally oral discourse that is usually accompanied and supported by gestures and can even be replaced by a system of visible gestures.

The practical world, the environment for elementary understanding, is a partial, dependent structure within the structure of a cultural lifeworld in general. This structural system of intentional objects as an environment given for elementary understanding can be called "natural" because it includes more than a "practical world" as a system of practical interactions. It includes the experience of disturbing and destructive powers that are always able to disrupt, to interrupt, and to destroy systems of goal-directed practical interactions. It is even present as the understanding of the frightening possibility of the complete destruction of all conditions of the practical world and elementary understanding (49; cf. 393 for "violent hostile interactions disrupting all civil interactions").

Seebohm next distinguishes "first-order" and "second-order" elementary understanding:

> The first-order understanding of the social interactions of the participants in the interactions can be understood in the second-order understanding by consociates and contemporaries. The whole context of first-order and second-order understanding is the empirical basis for the interpretations of social interactions and individual actions in social psychology and the empirical social sciences. First-order elementary understanding is, with a grain of salt, roughly equivalent to Weber's and Schutz's subjective or insider understanding of the meaning of an action or an interaction. Second-order elementary understanding is Weber's and Schutz's "objective" or outsider meaning. (310)

Seebohm here accepts my proposed alternatives of "outsider" and "insider" for Schutz's "objective" and "subjective" that can be applied to the participants in interactions and any observer[9] of them in everyday life as well as to any cultural scientists

experience on the level of animalic understanding of one's own body and its relation to the surrounding bodies" (220). There is counting (191) and measuring (217). Measuring can use a measuring rod (192), reckoning can use the abacus (192, cf. 343, 334), but the hour glass is not mentioned. Then again, arithmetical and geometrical techniques can be applied (266, cf. 329, 331), there is also "the mechanical art, the *mechanike techne* as the art of producing instruments, i.e., tools, and machines in the crafts" (214), medical arts (241), and finally "weapons, i.e., 'tools' that can serve as instruments in social interactions" (361), which presumably include war and crime as well as law enforcement.

[9] To refer to the distinction between the subjective meaning of an action and its social context on the one hand and their objective meaning for an observer on the other is misleading. An onlooker can be, and usually is, involved in the context of acts in interactions. The "objectivity" is restricted by concrete practical and social interests. It is not very helpful to add that the onlooker or observer ought to be disinterested in the social actions of the actor. Without fu1iher specifications such an onlooker would be either guided by idle curiosity or restricted to an empty stare. Scientific observations are interested. Their interest in the human sciences is precisely the objective validity of the interpretation of the action or life expression. But this explication of the meaning of a "disinterested onlooker or observer" is obviously circular. The questioning of participants in social interac-

and philosophers at higher levels. He also discusses "subjective meaning" (282, 289, 319–321) and Schulz's later synonym, "subjective interpretation" (319). We will return to this distinction.

There is one word that Seebohm seems to take from Schutz with confusion, namely, "contemporary." It is clear enough from the context when the reference is simply to subjects living at the same time regardless of whether they share space or cultures when the contemporary texts of a given text are referred to (e.g., 109, 112, 309). Sometimes when Schutz's terminology is referenced, it has a broad signification that includes consociates, something that Schutz does at least once, and then other times it has the strict connotation by which consociates are distinguished from contemporaries. If we recognize this difference, Seebohm is correct to say that, "Consociates are contemporaries in one's own immediate cultural lifeworld. Contemporaries outside this context are absent, foreign to one's own lifeworld, to different degrees (53).

Consociates, for Schutz, immediately share space as well as time, while contemporaries are only alive at the same time as the subjects. Then there is only indirect communication, and each has only fixed life expressions from the other.[10] Thus, this passage is mistaken: "The region of empirical material for the systematic human sciences includes … fixed expressions of contemporaries [sic] and immediate life expressions of contemporaries that are accessible for intersensory observations in the present" (404). One might think that this is merely a slip of the pen whereby "consociates" was intended, but then again there is this: "The objects in the empirical region of systematic human sciences, i.e., psychology and the social sciences, are life expressions of contemporaries given in the present including fixed life expressions and immediate life expressions" (91). In Seebohm's defense, one can contend that he intended that immediate life expressions were *possible* for contemporaries when they are brought into the consocial situation, for he does say that "consociates and contemporaries and their actions and interactions are given in actual or at least possible observations as immediate life expressions" (310). Other misleading uses of "contemporary" will be seen below.

---

tions causes, however, serious difficulties for observers who want to be in *disinterested* contemporary observers in every respect. It was and is a generally accepted methodological principle of the systematic human sciences that observations of social interactions and the secondary understanding of social interactions ought not to be involved or interested in the social interactions. The difficulty given with this methodological requirement is that research in the social sciences and in social psychology needs communication between the researcher and the objects of research" (Seebohm 2015, 289).

[10] Seebohm recognizes fixed life expressions besides texts traditionally understood: "In the last two centuries technology has offered ways of producing new types of fixed life expressions beginning with photographs and records to videotapes; such inventions cause additional methodological difficulties for the systematic human sciences and for contemporary history. It is easy to manipulate such "sources" for attempts to reconstruct "what is or was really the case." Required are new methods of "critique" for these new types of "historical" sources and these methods will have to use results of the *natural* sciences" (Seebohm 2015, 291 n. 29). I might add that he declined to use email and probably hardly knew of texting and Skype, which of course can also yield fixed life expressions.

Less importantly, I believe that for Schutz "successors" are others who continue after the time the subject begun from has died, while for Seebohm, "Some successors were known in the present as younger members of the social lifeworld" (52); he should have said "some contemporaries would become successors."

Finally, there is this handsome passage that not only summarizes much of what is said above but nicely leads to the next section of this study.

> Several types of understanding and life expressions and each with its significance for different aspects of the lifeworld, have been distinguished: animalic understanding, elementary understanding, and first-order higher understanding and second-order higher understanding. First-order higher understanding is the *creative* understanding of the context of the lifeworld as a whole and of partial aspects of the lifeworld. Such aspects include structures of political power, of custom and law, of trade relations, of intercultural contacts, of the production of goods, and last, but not least, of the natural environment. There is, on the other hand, second-order higher understanding, the *re-creative* understanding of the manifestations of first order creative understanding in immediate and fixed life expressions, the *interpretation* of life expressions.
>
> Seen from the viewpoint of these distinctions, the human sciences are not sciences practicing elementary understanding or first-order higher understanding. If this were the case the human scientist would be a prophet or a poet or a lawgiver or a natural scientist. These all create different types of first-order higher understanding. The understanding practiced in the human sciences is re-creative second-order higher understanding. They are "interpreters" of the life expressions of others. (88,154)

7.3 Three Specific Abstract Reductions

In my terminology, the issue here is that of when and how the cultural sciences become theoretical. Phenomenologically, my contention is that there is an initial *epoché* by which the everyday lifeworld is purified of the values and purposes it had for just the cultural scientist herself before her attitude was reduced from the everyday to the theoretical. My *philological* contention, however, is that Seebohm actually does recognize such an *epoché* even though he would prefer to call it an "abstractive reduction" and the attitude "disinterested." And because he sees difficulties in adopting and maintaining that attitude, he fails to emphasize this third specific reduction.

That he at least once recognizes a third abstractive reduction is clear:

> The human sciences are, according to Dilthey, but also others, e.g., Alfred Schutz, understanding sciences. Presupposing what has been mentioned above about the typology of understanding, it can be said that within the lifeworld, the human sciences are only interested in and reduced to the region of *secondary understanding* of first-order elementary and higher understanding of *Others,* i.e., the understanding of the meaning of immediate and fixed life expressions of contemporaries and predecessors as observable objects. All other aspects of the lifeworld given in first-order elementary with higher-order understanding of the interpreter are excluded. *This abstractive reduction that determines the region of the human sciences in general* can also be understood as a partial explication of the meaning of "being a disinterested observer" in the human sciences. (Seebohm 2015, 94, emphasis added).

Due to shortage of space, contemporary history,[11] ideal types,[12] onlookers, verification, and "methodological abstraction" will be omitted here.

Given my terminological preference, I began by looking for usages of "theory/ies" and related words and found "theory" occurring on practically every page and thus too many to study. "Theoretical attitude" occurs only four times, three of which refer to the sciences, e.g., "the theoretical attitude of philological and historical research" (351, cf. 169, 213), and one affirming the "distinction between the theoretical attitude of a philosophical ontology and the practical attitude of elementary understanding" (266). Then again, political science is said to be a "theoretical empirical science" (373, cf. 352) and "theoretical psychology" (300) is said to conduct its interpretations in the "theoretical cognitive attitude" (298). "Theoretical interest" occurs a dozen times. We are told that, "The cognitive attitude in the natural sciences is theoretical" (185, cf. 93) and "Research in the empirical systematic human sciences does indeed require a theoretical interest in actions and interactions" (293).

Historically, at least since Aristotle, "Pure contemplative theoretical understanding is the business of philosophy, including the "pre-scientific ontology of nature"

---

[11] "The basic epistemological problem for the systematic human sciences is that this distance between the present and past historical reality shrinks in contemporary history and reaches the zero limit for the interest in the future development of social structures given in the actual present in the social sciences. The type of distance that is left for possible "value-free" objectively valid research in the systematic human sciences is the distance between a *disinterested* observer and interpreter of Others in the space of systems of social interactions in the present lifeworld. The objects of research are Others who are participants in social interactions and their first order understanding of the social interactions. But these "objects" are themselves other subjects who can participate in communications and interactions with the subjects carrying out the research. As such they are potential partners in dialogues in the present. In contrast, the predecessors who are the authors of the presently available fixed life expressions, i.e., the immediate objects of historical research, cannot appear as partners in dialogues in the present. Research in the systematic human sciences, i.e., psychology and the social sciences, is first of all theoretically "interested" in events in the *present*" (Seebohm 2015, 411, cf. 95).

[12] "According to Schutz, it is possible to use the term "ideal type" for certain essential dependent or independent partial structures of "course of action types." Schutz distinguishes (1) course of action types; (2) personal ideal types guided by typical in-order-to-motives; (3) cultural products or artifacts (including tools). For Schutz type (2) can be derived from type (1) and vice versa.

Type (3) can be understood the realized product of (1) as guided by in-order-to-motives. The personal ideal types (2) are the ideal types or constructs that can be understood as puppets or *homunculi* in the classification of Schutz. A precise understanding of this classification has to cope with difficulties. The main difficulty is that there seems to be a one-to-one correlation between a course of action type and a personal type. It follows that it will be difficult for Schutz to distinguish clearly between a phenomenological epistemology *(Wissenschaftslehre)* for the *social* sciences and a phenomenological psychology serving as a phenomenological epistemology for social psychology as an empirical science. It cannot be the task of a systematic investigation to decide problems of the interpretation of Schutz's theory of ideal types and constructs and his division of different types and constructs. What had to be emphasized is only that his theory is, though partially modified and extended here, the presupposition of a formal and material analysis of a system of different subtypes of the general type of ideal types that can be applied in the social sciences" (Seebohm 2015, 315).

(174) and the "philosophical ontologies after the emergence of *episteme*, i.e., pure theories. Their cognitive attitude includes the abstractive exclusion of all practical activities" (273). This exclusion may hold potentially but not actually for all theoretical science:

> The thematic attitude of the sciences is interested in the *active* search for causal relations and the different types of matter presupposed in causal relations. It shares the interest of elementary understanding in the discovery of causal connections without being restricted in this search by the interest in the purposes of practical life. And … this interest of the thematic attitude of the sciences is theoretical, but it is not contemplative. It is active because it implies the actions in experiments or the hunt for opportunities for observations that can confirm or disappoint predictions, confirm or falsify hypotheses. (2015, 175)

Otherwise, "contemplative" occurs on 12 pages, two pertaining to political science (352, 374) and three pertaining to Heidegger and metaphysics (384, 390, 394). The rest are pertinent here. Above all, there is the higher understanding to which science specifically belongs: "Elementary understanding will be characterized as the understanding at work in the interactions of practical life and higher understanding as the contemplative theoretical understanding of the lifeworld, of the human condition in religion, and philosophy, and last, but not least, the sciences" (6).[13]

As for "disinterested," there are 48 occurrences in the new book. In the earlier book, there is one occurrence, which is interesting but not relevant here.[14] It is the key word for understanding the abstractive reduction for the human sciences and, "'Disinterested' means to be disinterested in activities that can cause changes in the course of events in the region of these objects" (410).

There is no question that both species of the natural sciences are or can be disinterested or, in my terms, theoretical.

[13] "Higher understanding is contemplative. It presupposes time that is not dominated by the needs of the practical activities of elementary understanding. Of interest for higher understanding are all relevant aspects of elementary understanding. Of interest for higher understanding are, furthermore, the social structures of the lifeworld and changes in the social structures, including customs such as fashions and styles, as well as customary laws and written laws with their distinctions between right and wrong in interactions. Of interest are, moreover, significant deeds of members of the community that have changed structures of the lifeworld in the past, significant changes in the natural environment of the lifeworld, and encounters with foreign lifeworlds. The manifestations of higher understanding are cults, artifacts serving cult activities, myths, prophetic revelations, poetry, philosophical reflections, and finally the sciences" (Seebohm 2015, 51).

[14] "There are several misunderstandings concerning the type of objectivity governing both philological-historical research and archaeological research. The first is implied in the claim that pure and disinterested contemplation governs the attitude of the researcher. Such a claim was made in the nineteenth century. In a certain sense this is true, but seen from another viewpoint it is an illusion, because it can be said that the present in general transcends the past. It considers objects in the past as objects at a temporal distance. But it is already obvious for subjective time-consciousness that every present will flow off into the past. Pure contemplation of the past is possible, but it is possible only in the present. Moreover, further reflection can show that the contemplation was not pure after all, because it was determined by the temporal situatedness of the contemplator, and further reflection is always possible if the contemplation and its life expressions has flowed off in the past" (Seebohm 2004, 248).

> Observations in the empirical natural sciences can be recognized as "disinterested observations" precisely because the natural sciences are theoretical *empirical* sciences. They are *disinterested theoretical observations* only because the residuum of the *abstractive reduction* brackets precisely the categorial structures that are characteristic for practical interests in general and therefore also for interested observations. It is, furthermore, meaningless for the ontological region of the natural sciences to assume that natural scientists can be tempted to ask their objects (e.g., planets, minerals, plants, magnetic waves, etc.) whether they agree or disagree with the scientists' understanding of their interactions, e.g., in the solar system. (2015, 294)

Furthermore, as stated, with adjustments, in the epigraph above:

> The ... reduction [which leads to the biological sciences] abstracts from the purposes and values of everyday practical life, and in general from all contents that are given in lived experience and reflections on lived experience. The ... abstractive reduction [which leads to the physical sciences] abstracts from secondary qualities of observable objects that are given in the residuum of the first abstraction, reducing them to primary qualities.

As for the human sciences, there are two subspecies of this species, the historical and the systematic, the latter including psychology and the social sciences in particular.

Concerning the historical sciences, which have an abstractive reduction (284):

> The attitude of historical research is the attitude of a "disinterested observation." This means in this case that the meaning and significance of the fixed life expressions is given as meaning and significance for Others at a temporal distance. There is no immediate need to apply or to reject them in the own context of one's own tradition in the present. (2015, 152)

Seebohm also includes philology and archelogy as historical human sciences (72).[15]

The problems for Seebohm are with the systematic human sciences. He is concerned with how an investigator or researcher can distort data, alter social reality, and thus invalidate accounts. She seeks to understand data that are expressions of the "object's" or, better, the informant's "subjective interpretations" (or "subjective meaning"), for which he accepts Schutz's concept:

> It presupposes that the actors know where their action starts and where it ends, and what their "in-order-to-motives" and their "because-motives" are. In terms of the typology of understanding laid out in this investigation, actors have a first-order understanding of the purposes and the efficient means used for the realization of the purposes in their actions. In social life, the individual actor's knowledge of the action also implies that this action is given as an aspect of an interaction in a shared intersubjective world, i.e., in a structural reciprocity of perspectives. This subjective meaning can be understood by the partners of the actors who are also involved with the actors in the interaction because they share a set of purposes with these actors. Moreover, it is understood, and this means it can be understood in *secondary understanding* ... by consociates and contemporaries [sic] who are not immediately involved in the interaction.

[15] "Modern historical research presupposes in addition archaeological interpretations of presently given aliifacts and monuments created in a past lifeworld. Archaeological interpretations can use interpretations of texts with information about the use and the purpose, i.e., the meaning, of artifacts and monuments whenever this is possible. This possibility is, however, restricted to historical archaeology. The situation is different in pre-historic archaeology, i.e., in the case of past lifeworlds without literary traditions!' (Seebohm 2015, 102).

> Interpretations of the subjective meaning of social interactions for the participants, the actors in such interactions, require in addition a secondary understanding that is not only not involved in, but disinterested in the interaction. (2015, 319)

The original situation is for consociates, and the fundamental data are expressions by actors. Onlooking observers and contemporaries are secondary. Also, immediate life expressions are given, e.g., posture, visage, gesture, intonation, and pacing of speech, etc., which Seebohm alludes to but does not emphasize them. Where the establishment of disciplines is concerned:

> The empirical basis of the systematic human sciences is not a specific region of objects given in the concrete lifeworld that can be determined by an isolating abstraction. concrete lifeworld as a whole. The disciplines of these social sciences are different because they are interested in different aspects and not in separable regions of the lifeworld. (2015, 284)

References

Embree, Lester. 2015. *The Schutzian theory of the cultural sciences*. Dordrecht: Springer.
Seebohm, Thomas M. 2004. *Hermeneutics: Method and methodology*. Dordrecht: Springer.
———. 2015. *History as a science and the system of the sciences: Phenomenological investigations*. New York: Springer.

Lester Embree (New School, 1972) studied with Edward G. Ballard, Dorion Cairns, and Aron Gurwitsch. He taught at Northern Illinois University, Duquesne University, and now Florida Atlantic University, where he was the William F. Dietrich Eminent Scholar in Philosophy until his death in 2017. From 1985 to 2005 he was President of the Center for Advanced Research in Phenomenology, and he led the founding and early development of the Organization of Phenomenological Organizations. He has translated works of Suzanne Bachelard and Paul Ricoeur; edited work of Cairns, Gurwitsch, and Alfred Schutz and also various collective volumes, the *Encyclopedia of Phenomenology* (Kluwer 1997) included; and authored a number of essays in and on constitutive phenomenology, as well as *Reflective Analysis*, second edition (2011). His deepest interest is in the theory of the cultural disciplines, the theory of American theoretical archaeology in particular.

Chapter 8
Mathesis and Lifeworld: Some Remarks on Thomas Seebohm's *History as a Science and the System of the Sciences*

James Dodd

Abstract This essay draws inspiration from Thomas Seebohm's remarks on the origins of mathematical thinking in the lifeworld in his *History as a Science and the System of Sciences*. It argues that the crux of the phenomenological account of the origins of mathesmatics lies in the temporal constitution of mathematical idealities, above all with regard to phenomena associated with measurement. Such an account in turn promises to provide a sound basis for a nuanced conception of mathematical thinking that is well-suited to articulating problems germane to modern mathematics and the philosophy of mathematics.

The nineteenth century mathematician Leopold Kronecker is famous for several important ideas and discoveries in algebraic number theory. He is also famous for expressing something of a curmudgeonly resistance to the revolutionary winds of contemporary mathematics, memorably summed up in the quip "The whole numbers were created by the good Lord, everything else is the work of men."[1] In most accounts of the history of mathematics, the view Kronecker expresses in such statements, even if it was certainly held by others, is rarely characterized as more than a brief, quixotic objection to the revolution being carried out by Karl Weierstrass, Georg Cantor, and Richard Dedekind, among others, and his curt remark is often reproduced to elicit a knowing smile of affection for the hopelessly stubborn among us.

[1] "Die ganzen Zahlen hat der liebe Gott gemacht, alles andere ist Menschenwerk." The mathematician Heinrich Weber attributes this to Kronecker in his eulogy "Leopold Kronecker" (1891/1892, 19).

J. Dodd (✉)
New School for Social Research, New York, NY, USA
e-mail: doddj@newschool.edu

© Springer Nature Switzerland AG 2020
T. Nenon (ed.), *Thomas Seebohm on the Foundations of the Sciences*,
Contributions to Phenomenology 105,
https://doi.org/10.1007/978-3-030-23661-8_8

Thus I was surprised to see Thomas Seebohm apparently embracing the spirit of Kronecker in a discussion of mathematics in his book, *History as a Science*:

> A phenomenological version of Kronecker's famous dictum "Whole numbers have we by God, everything else is the work of men" is "Natural numbers and basic arithmetical *operations* are given in the lifeworld for elementary understanding, everything else is the work of mathematics as a science belonging to higher understanding." (2015, 190)

Yet this embrace of Kronecker's inveterate conservatism is only apparent. Kronecker was objecting to the willingness of mathematicians to follow innovations in ever more abstract methods of mathematical analysis, wherever they may go, thus risking the loss of our most immediate and basic intuitions regarding the meaning of number, or the objects of mathematics generally. Seebohm has something else, and arguably more nuanced, in mind with his appropriation of Kronecker's dictum. His main goal is to emphasize the role of the lifeworld in the origins of mathematical thinking, and he is motivated not so much by the need to defend a given, putatively more "intuitive" conception of number, as to illuminate the complex origins of any concept of number of whatever kind and on whatever level of abstraction.

To spell this out more precisely requires first a look at the typology of understanding alluded to in the quote above, which plays an important systematic role in Seebohm's book. The key distinction is between what Seebohm describes as an "elementary understanding," or an understanding defined as a "system of goal-directed interactions" operative in the environment or *Umwelt*, and what he calls a "higher understanding," which is embodied in the attempt to make sense, or to clarify the world in a way that strengthens and fortifies the success of elementary understanding as a fundamental orientation to and in the world. Both dimensions of understanding are intersubjective, rooted in interpersonal experiences of things and persons, and shaped by practices both contemporary and historical—in short, both are forms of understanding germane to the *Lebenswelt*, and make up its basic epistemic as well as practical profile.

One of Seebohm's most interesting suggestions is that the contrast between these two levels of understanding turns on the experience of *disruption* and *breakdown* on the level of elementary understanding. Disruption, on Seebohm's account, gives rise to higher understanding as a response, one that seeks to re-establish the stable comprehension of things originally enjoyed by elementary understanding. This theme of disruption is an important part of Seebohm's account of the generative origins of science that, in my view, has potentially significant consequences, especially when we consider its role in Seebohm's account of the genesis of mathematics.

It is worth quoting a couple of the passages in which Seebohm makes this contrast explicit:

> It [the environment given for elementary understanding] includes the experiences of disturbing and destructive powers that are always able to disrupt, to interrupt, and to destroy systems of goal directed practical intentions. It is even present in the frightening possibility of the complete destruction of the conditions of the practical world and elementary understanding. The first and main task of higher understanding is "to make sense," to understand and explain the blind forces, human forces included, behind the natural environment as a correlate of elementary understanding. (2015, 51)

> Elementary understanding is blind without higher understanding, but higher understanding is empty without elementary understanding. (2015, 52)

Higher understanding thus seeks to rearticulate a given comprehension of the world, one already embedded in practical dealings with things, in a way that secures it against the potential threat of disruption, or, better, the recurrence of a disruption of a kind already suffered. It is the attempt to respond to the experience of incomprehension with illumination and, with that, a renewed and thereby more reliable effectiveness in pursuing our ends. Such response is broadly determined: higher understanding for Seebohm comprises a range of different activities and cultural formations, including "cults, artefacts serving cult activities, myths, prophetic revelations, poetry, philosophical reflection, and finally the sciences." (2015, 51–52)

The focus of *History as a Science* is, of course, science. Science for Seebohm is essentially the complex activity of formulating a "causal explanation" of things, one that has its generative origin in a pre-scientific lifeworld as one way among others to make sense of disruptions in our understanding that frustrate our capacity to navigate our world. Causality, taken in a broad sense, is a very primitive concept, one that operates quite independently of science, and in sophisticated if not always compatible ways. In Seebohm's terminology, the theme of causality belongs to elementary understanding, though of course once it is taken up in higher understanding it undergoes considerable refinement (and often, arguably, reification). More specifically, Seebohm understands the emergence of natural science from the pre-scientific lifeworld in terms of an *idealization* of causality or, better, its re-articulation in purely *mathematical* terms.

There are thus at least two fundamental questions that guide this analysis of the emergence of science (or, to use Seebohm's apt phrase, "lifeworlds with science"). The first is the question of the role of the primitive meaning of causality in elementary understanding, or how an orientation to causal explanations in the broadest sense is generated in the context of the lifeworld. The second question (or questions) revolves around issues of the conditions for the possibility of an orientation to causal explanation taking on the explicit form of a *mathematical* idealization. This roughly follows the itinerary of Husserl's *Origin of Geometry*: the first thread of analysis is a kind of archaeological exercise that seeks to identify the most primitive forms of relational consciousness in which the lifeworld is constituted at its most basic level; the second thread, building on the first, seeks to identify the sources for the proto-accomplishment of a consciousness of ideality within the horizon of experience (Husserl 1954).

It is this second strand of analysis that I wish to comment on in what follows, or at least on some of its basic elements. The essay will take the following form: First, I will consider Seebohm's account of the origin of mathematics in elementary understanding within the lifeworld, the central topic of which turns out to be *measurement* (or the sense of the measurability of things). Next, I will point to an interesting, though relatively implicit, suggestion of Seebohm that primitive measurement is grounded in potentialities that belong to our experience of time, or the basic structures of inner time-consciousness, to put it in more phenomenological terms. I will then conclude with some critical observations, as well as some suggestions regarding how Seebohm's approach to understanding the rootedness of *mathesis*, or mathematical thinking generally, in the lifeworld might be extended for its own sake, beyond the immediate concerns of *History as a Science* and into a more general phenomenological account of the experiential origins of mathematical thinking

As I have already intimated, key to Seebohm's account of the natural sciences as an instance of higher understanding is the origin and generation of the idea of the measurability of things, or the possibility of establishing a perspective on the surrounding world in which phenomena "are relevant only as measurable observables" (2015, 167). The establishment of the *measurability of the given* in turn provides the basis for the mathematical representation of the given and its relations as such, and by extension the world; it provides, so to speak, the basis for a general ontology of natural existence, even if it can also remain implicit in a mere instrumental perspective oriented to the success or failure of predictions.

The idea of measurability takes hold, in part, due to the success of what Seebohm describes as the interweaving of two abstractions. The first is the abstraction from the immediate practical interests of elementary understanding that establishes a properly theoretical, in the sense of relatively value neutral perspective. Seebohm does not use this language, but the idea of the neutralization of practical values is implicit in the following gloss of the contrast between the interests of the one and the other level of understanding: "It [the thematic interest of science] shares the interest of elementary understanding in the discovery of causal connections without being restricted in this search by the interest in the purposes of practical life." (2015, 175)

One might quibble here a bit and suggest that what Seebohm has in mind is more of a *modification* than a proper *abstraction*: after all, an interest in causality is an interest in causality; the difference here comes down to the modified interest of the theoretician to clarify the relevant causal relations beyond what is immediately demanded by a given practical interest, instead of being satisfied with the level of clarity that is sufficient to achieve a given purpose. At bottom this is not so much the abstraction of an operation of clarification from its natural home in practical understanding, as it is a modification of practical understanding into an explicit, higher order practice of clarification.

I suspect this slight reformulation might in turn help make stronger Seebohm's important point that what distinguishes science from other forms of higher understanding is that its development has a material impact on elementary understanding and its environment, specifically in the form of technological advancement. The inner relation between science and technology can only become evident if we provide a nuanced enough contrast between theory and practice, which Seebohm repeatedly does by emphasizing that science must be understood above all as a *practice*, presumably *pace* its "abstraction" from the "purposes of practical life."

The second abstraction Seebohm posits, and which he understands as having been made possible by the first, is more unambiguously an abstraction: it involves the separation of all that is measurable from anything that is not measurable, from any quality or feature of things that cannot be captured by a measure. The two abstractions work together, though in somewhat different directions: while the first abstraction (or modification) extends the horizon of reflection in which causal relations are both discovered and articulated, the second reduces the diversity of phenomena into something slightly more manageable, admitting nothing into view apart from what is directly or indirectly measurable.

This second abstraction is possible, for Seebohm, only given the accomplishments on the level of elementary understanding that have already put into place a fully developed arithmetic (even algebra) and geometry, not so much in the sense of "pure" sciences, as a knowledge of number and form embodied in the practical art of measurement. Yet the second abstraction also relies, as Seebohm goes on to argue, on the presupposition of a specifically *ontological* mode of reflection having already taken hold in the lifeworld, as he puts it in the following passage:

> Necessary generative foundations for the emergence of the natural sciences in a cultural lifeworld are in addition to a sufficiently developed geometry together with a sufficiently developed algebra and a philosophy of nature, a contemplative reflection on the ontological structure of nature in general. (2015, 181)

The suggestion here seems to be that, without this ontological impetus, so to speak, mathematics would remain relatively inert with respect to the explanatory demands of higher understanding, even if it becomes more articulated in its purity. One might also suggest, though Seebohm does not make this explicit as a problem, that not only does the mathematical science depend on an ontologically oriented reflection, but equally mathematics *as* science.

Though the question of the origin of an ontological perspective (or more generally a "philosophy of nature") is in many ways decisive for the argument, my focus here will be on Seebohm's analysis of how the pre-scientific lifeworld establishes a sense for the measurability of things, with or without ontological reflection, and thus provides a place for a *mathesis* that can in turn form at least one of the generative conditions for the emergence of modern mathematical science, above all physics.

The phenomenological problem of measurability has two parts, on Seebohm's account, which together roughly amount to what Seebohm likens to a phenomenological justification for Kant's axioms of intuition in the first *Critique*.[2] This analogy with Kant's project is defensible in only an extended sense. The first part of Seebohm's approach is to give an account of the ground for the manifestation of mathematical objects and relations; this is the familiar phenomenological problem of the experiential constitution of mathematical objects, such as numbers and geometric figures, but in general it is arguably rather alien to Kant's understanding of mathematical objectivity. Kant's approach involves the identification of transcendental principles that underlie the constructability of mathematical objects in intuition in general and the measurability of appearances in particular. This is a rather

[2] The principle of which is "All intuitions are extensive magnitudes." See Kant 1902, A162/B202–A166/B207.

different perspective than phenomenological constitution, in which mathematical objectivity is not from the beginning reduced to subjective *a priori* principles, but instead formulated in terms of the manners of givenness germane to its objects. The second part, which is more congruent to Kant's conception of the axioms of intuition, is to give an account of how this givenness of mathematical objects relates to that of real objects in experience, whether on the level of elementary understanding itself, or within the bounds of the abstraction of all that is measurable. (2015, 186)

Seebohm's analysis is overall rather complicated, even before we get to the level of idealization, which is for him the terminus of the genetic development he is tracing. He begins with a clarification of the generative conditions for the givenness of mathematical objects, thanks to which things like numbers, or rather numbered collections, and geometrical forms become thematic. Only once we have clarified these generative conditions of the manifestation of mathematical objects, Seebohm argues, will we be able to tackle the second, related problem of the applicability of properly mathematical concepts to objects in general. And only then, once the phenomenological profile of mathematical objectivity and the patterns of conceptual applicability that constitute the measurability of things is in place, will be we able to tackle the idealizations that make up the foundations of science, both formal and material.

It is in the context of introducing this double (or maybe triple) problem that Seebohm evokes Kronecker. Yet again his point is not so much to make a contrast between the supposedly more "genuine" reality of the whole numbers against the fictional creations of abstract number theory, as to argue that the basic accomplishments of elementary understanding in making something like "number" manifest *at all* hold the clue as to how the extensions of the concept of number in the domain of higher mathematics and, taking measurement into consideration, mathematical sciences of nature, is at all possible. I don't take Seebohm, in other words, to be attempting to reign in excessive mathematical speculation; he is more interested in the universal generative structures of cognitive experience thanks to which mathematical objectivity emerges as an evolving field of both theoretical objectivities and their applications.

It is arguably important at this point to note that Seebohm's book is a systematic work on the scientific status of *history* and what he takes to be its mediating relation between the human and the natural sciences. It is not, in other words, a book on the foundations of mathematics proper. Accordingly, Seebohm presents only just as much as he needs to sketch the phenomenological origins of the mathematical character of natural science. He is thus often content to refer the reader to the work of others, such as Richard Tieszen and Dieter Lohmar, to fill in the lacunae.[3]

There is, however, at least one basic insight that Seebohm develops on the origins of mathematical thinking that is worth further reflection beyond the immediate hori-

[3] Seebohm cites Lohmar 1989 and Tieszen 1989. We can add the more recent volume edited by Mirja Hartimo (2010), which includes contributions from several scholars working on the phenomenological foundations of mathematics (and logic), including Guillermo Rosado-Haddock, Claire Ortiz Hill, Robert Hanna, and Olav Wiegand.

zons of *History as a Science*. It can be found in his contrast between two perspectives basic to the activity of *counting*, each of which Seebohm describes, significantly, in explicitly *temporal* terms, as for example in the following passage:

> Counting presupposes either the collecting of things one by one into a collection or a re-recollecting of a pre-given but not yet counted collection of things, starting with a deliberately chosen thing and adding other things one by one. Counting as an intentional act implies intentional objects that are given in two intentional perspectives. On the one hand, there is the presently given number of things in the already counted collection, the numerical quantity [*Anzahl*] of things in a finite *set* of things. On the other hand, the things are given in a temporal sequence of intentional acts starting in a past Now in the past horizon of the actual Now, and this temporal sequence determines and represents the sequential *ordering* of the counted *things* in the collection. (2015, 191)

This of course alludes to the notions of cardinality and ordinality respectively, but more important here is again the emphasis on temporality. The first perspective is described as taking shape along the lines of the *already having been formed* of a collection of counted things (where counting amounts to adding things one by one to the already formed collection, where the last counted represents the total number of things counted), thereby determining its presently given categorial formation as a "set"; the second perspective is described in terms of the horizon of collecting itself as an act of synthesis which, relying not only on the pre-givenness of what is to be counted but also on its own characteristic retentional depth, provides the basis for a properly sequential "one after the other" as a unity of sense or meaning that lies at the core of the concept of succession. Seebohm does not make this connection, but when he turns to the use of mathematical signs on the very same page, the meaning of "sign numbers" presupposes precisely this already constituted temporal density of either the ordinality of numbered objects in sequence or the cardinality of groups of counted things. Number signs, though they represent "anything whatsoever," as Seebohm emphasizes, and thus also mark the emergence of number as a generality, can only be an "anything whatsoever" that has always already become manifest in accordance with a definite temporal profile, or manner of passage at play in one or the other (or both) of the perspectives. To read a number sign, in other words, is an act of interpretation that has been originally primed by a specific manner of experiencing numbering in time, and the modes of objective manifestation that it has made possible.

Counting, however, is not yet measuring. In Seebohm's account, what we might call the complex temporal density thanks to which number becomes originally manifest again becomes significant when we turn to measurement as a special application of numbering. The measuring of *space*, or rather the filled space of intuitive appearance (Kant's "extensive magnitude" [*quantum*]), is perhaps the paradigmatic example, and Seebohm considers it first. A spatial interval itself, unlike the discrete objects that occupy such intervals, does not in fact lend itself to being "counted" in either of the two primitive senses described. An interval of space is first rendered countable (perhaps even manifest as an "interval" at all) only thanks to the delimitation of a unit against which a given stretch of space can be *compared*. The most primitive example of this is probably using one's hand or foot to pace out a given distance across a surface, or even just the *ground*, a semantic association that survives in the very term "geometry." Another primitive example would be employing a given object, such as a rod of a fixed size (*rigidity* being another essential component), to mark out a larger spatial interval.

And here, too, time plays a key role, providing the constitutive ground for the extension of the countable into the measurable. This role of time is, however, only implicit in Seebohm's analysis, so we must first tease it out.

To illuminate the role of time in the constitution of extensive magnitudes, first consider the following point Seebohm makes about the basic function of the measuring rod:

> The measuring rod is supposed to represent—not to be—a measured distance that is spatially not approximately equal but *exactly* equal to any other spatial unity of the size of the measuring rod. (2015, 192)

Seebohm does not detail what he means by the contrast between the rod *representing* a measured distance and something *being* a measured distance, but the intent seems to be clear. The emphasis in measuring is on the determination of equivalences, so, for example, how many units it takes to "equal" a given spatial interval. We thus do not count "space," but rather count an interval "in" numbers of units; for example, we mark out the path from the garden to the road "in" feet, or the length of the block of granite "in" yards. Standard units are themselves of course arbitrary, but once established they are comparable and, more importantly, inter-translatable. Builders coming from different parts of ancient Greece used a variety of different standard units of measure, but there is evidence that they developed a table they could use to translate instructions given in one standard into whatever one they were used to using (just as an American carpenter in Europe learns to habitually shift between the English and metric standards of measurement) (see Jones 2000). The important point is that any such unit, from our primitive rod to the meter, is not itself measured "in" the unit so defined; it can only itself be measured by *other* units, even if these are determined by simply subdividing the unit itself, thereby measuring the whole with one of its parts.

Yet this account raises a question—one that Seebohm does not raise, for again his treatment of these matters is very economical—as to the generative conditions of the establishment of such a unit of comparison in the first place. The rod is not itself measured but represents the measurement of anything equivalent to its size. It seems clear that it can do so only if the rod has already been established as the operational center, so to speak, of an act of comparison, one that moves beyond the mere contrast of "larger" and "smaller," and instead pulls the activity of measurement into the domain of counting. The question is what establishes the rod as a "measuring" rod proper, a *repeated* focus of comparison, and not just one object among others being compared with regard to the smaller or the larger? How, in other words, does magnitude become a properly mathematical phenomenon?

Seebohm gives us the negative answer: the unit of measure is not established as something *itself* measured, so it does not stand as the first measured space, the accomplishment of which would then somehow be extended to the measurement of other like or unlike "spaces." The rod is instead the accomplishment of the functional representation of the measurement of things of *equal size*. Thus the question for Seebohm seems to turn on the role of representation in the constitution of extensive magnitudes, but again he only provides a negative answer to the question formulated in this way.

I would suggest that the positive answer is in fact implicit in Seebohm's description of the two perspectives that are basic to counting that I described above, for the emergence of a representation of measure (a standard unit) arguably involves similar temporal structures that that are constitutive of both perspectives. This can be brought out if we consider that we can, in fact, ask a similar question about the role of units (the "ones" in the "one after another") in counting that we are asking about the role of a unit of measure (the rod): namely, how do we establish the sense-unity of a "one among others" that is more than the brute differentiation of individuated somethings, to an order of comparability that forms the ground of the specific form of composition basic to "a given number of x"? More concretely, what is the genesis of the function of a bead on an abacus, to evoke one of Seebohm's recurring examples?

A phenomenological reflection on these issues might accordingly go something like the following. It is clear that merely taking individual things together in some sort of vague aggregation based solely on their individuated differentiation falls short of counting; I can glance at the whole of an apple tree with "all" its fruit without apprehending the given aggregate as a "number" of apples. The perception of a heap is not the categorial intuition of a set; a trickle of increase does not an ordered series make. The individual thing alone, even apprehended in the unity of "this and another," is at most the categorial basis for the "this one and another one," necessary to yield an apprehension of the aggregate as a "number" of apples. For counting to be achieved, something else needs to be in play: an individual apple must be established as the *one* to which the others are *added*; it must be set up, much like the measuring rod, as a kind of relative, even if a temporary *marker* against which any other apple is to be compared—though now not in terms of extraneous comparability, such as size, but rather in terms of being just as much a "one" as any other apple, or for that matter any other thing. We might here add that this function of taking a thing as a relative marker in the act of counting relates to the sense of the assignability of number signs, or that the sense of the number sign as denoting "anything whatsoever," emphasized by Seebohm, already sets into play a primitive form of correspondence suggested by the marking of "one, then another *one*."

This is where the theme of time or the temporality of the act of counting might again make itself felt. For clearly the "one" must be set up as a marker specifically in *time*: it must be set up as an already *given* mark in the case of collection, of individual things that have already been marked by the operation of comparison (whether it takes the additional form of an addition or subtraction), and already thus "counted." But this temporal form of *having been counted* is in turn clearly dependent on the movement of counting itself, which arguably just is the generation of a sequence of comparisons out of the originally marked "first" element in the sequence held in retentional unity, to which the present counted "one" stands in a renewed, specifically oriented relation of a counted "and"—whether this "and" sets it into an order or just into a given collection retained in counting. Here lies the analogy of the operational center of the marked "one" in counting, understood in terms of its specific temporality, with the rod as a unit of measure: having been established, the rod as well can only function as a present unit of measurement in the form of the *already having been established* as the unit "in" which we measure, as the basic *mark of distance* (or in Seebohm's language "representation") that we use to "count" space.

There are two things to notice about this potential line of reflection. The first is that it suggests that, at least from a generative point of view, there is something of a primacy of ordinality when it comes to understanding our consciousness of number. Ordinality seems to cleave closer to how numbering unfolds in experience as a temporally coherent act, whereas cardinality seems to be either a less articulated form of ordinality or a kind of basic disassociation (or perhaps even abstraction) of partial results of counting from the process of the "one after the other." We should, however, be very cautious in such speculations. The aim of constitutional analysis is not to decide whether ordinality is *ontologically* prior to cardinality, as if phenomenology alone could decide theoretical questions that properly belong to mathematical reflection; instead the point is to delimit the resources inherent to mathematical thinking that allow not only for a rigorous distinction between ordinality and cardinality, but for the rigorous posing of precisely the question of their formal or ontological relation. Seebohm's own presentation of the issue suggests strongly that, from a phenomenological point of view, the perspectives of ordinality and cardinality are equi-primordial; this would mean that any apparent priority of ordinality is likely only a function of the fact that the common ground of time-consciousness underlying both perspectives is at best more perspicuous in the case of ordinality, given the tendency to think of time as inherently successive. This apparent primacy of ordinality from the perspective of the constitution of our consciousness of number would not be contradicted if, on ontological grounds, we were to conclude that cardinality in fact expresses the meaning of number in a more rigorous, comprehensive way—again, phenomenology cannot decide the matter. At most—and this is far from a modest contribution to the philosophy of mathematics—it can illuminate the inherent resources of mathematical thinking that allow for a progressive development from the grasp of "natural" numbers to more logically complex, sophisticated representations of number and numerical operations.

The second thing to notice is that, in both the cases of counting and measuring, what I above described as the operational center of equivalences, or the "unit," this center itself is neither counted nor measured, but only *established*. One might even say that it is *constructed*, and even in a sense that is evocative of a basic feature of traditional geometry that is both preserved and systematized in Euclid. Reasoning in traditional geometry follows the lead of constructions that serve to illustrate geometrical relations; the ruler and compass that such constructions deploy in turn follow the lead of the construction of basic temporal markers that originally serve to map out the very measurability of things. This suggests that, basic to the primitive phenomenological constitution of mathematical thinking, there may be a basic relation formed between construction and what constructions bring into the purview of mathematical intuition, without the latter necessarily being reducible to the former, or vice versa. The temporality basic to the constitution of mathematical objects, in other words, may allow not only for the basic constitution of points of reference and comparison, but also for the fluidity of construction, thanks to which mathematical objectivity opens as a horizon of discovery, one pursued through operations of re-orientation and re-thematization offered by constructions. If higher understanding, in the sense of Seebohm, involves the development of the accomplishments of arithmetic and geometry through their idealizations, it is perhaps the idealization of such basic proto-forms of construction that prove to be the most significant in any reflection on the conditions of possibility of modern mathematics, given the central importance of the role of formal axiomatization since the nineteenth century.

Still, one should note that such remarks—which are of course very tentative and speculative—are not of much importance for Seebohm's overall project. More important for him is the recognition that at this level of intersubjective experience—so on the level of what he calls elementary understanding and the lifeworld—we have the basis for the development of a focus on and interest in *precision*. For the fact that measurability comes down to a comparison of equivalencies in turn implies the conception of a more or less *exact* equivalence, so, for example, a precise mapping of a given standard onto a portion of space (or whatever is to be "measured"). This perhaps points to another connection to the figure of construction, if by "construction" we mean something that implies a broad sense of the making and fashioning of things. In this way, the full genealogy of properly mathematical construction includes the practical pursuit, say in carpentry, of the production of ever "more even" surfaces, "more regular" shapes, or "more symmetrical" patterns in the making of artifacts; likewise the development of more regular and precise tools for measurement as such, like when we discover that our trusty wooden measuring rod tends to swell and shrink under varying weather conditions, and we opt for a more reliably rigid metal rod instead. Or, to evoke an example from ancient Egypt, we treat our knotted measuring ropes with beeswax to prevent them from shrinking (or expanding). Such examples also point to how measurement inevitably interweaves with a growing understanding—again for Seebohm all of this is on the level of elementary understanding—of the behavior of different materials under different conditions, a body of knowledge that is arguably not unimportant in the empirical history of idealization.

The theme of construction also cuts another way, however. Seebohm is keen to emphasize the importance of approaching mathematical thinking as a set of *operations*, or more generally a practice, a doing, even a making. This again belies an easy differentiation of mathematical science from other forms of practical activity that involve mathematics. What is unique about the results of the kind of "making" germane to mathematical thinking, namely the manifestation of mathematical objects, is that their objectivity can be thematized by a turn away from the structures constitutive of such objects to a focus on the mathematical operations of the thinking that "produced" them. An important step in this direction is the development of an algebra that makes extensive use of variables for numbers and, once applied to measurement, for distances, thus shifting the meaning of mathematical form from being exclusively about mathematical objects to include operations with such objects, thereby opening the way (eventually) for the formalization of not only mathematically determined abstract multiplicities and their relations, but the theory forms of mathematical thought as such, which stand as their correlates.

What Seebohm has in mind here is of course the origins of modern axiomatic theory, but also the theory of vector spaces from Descartes to Hilbert, which rely heavily on abstract algebra and methods of formalization. He does not speak so much of "construction" as "formalism," but I think he has much the same in mind, though curiously he has little to say about the important role played by formal (or perhaps better formalized) logic in the development of modern mathematics. One need not be a logicist to recognize that much of modern mathematics emerges out of a turn to bringing mathematical *thinking* into view, resulting in much more systematic reliance on the laws of logic, even if (and when) this falls short of a full logicist program in the foundations of mathematics.

Naturally, in commenting on an impressive a work as *History as a Science*, which is evidently the fruit of a lifetime of reflection, one yearns for the opportunity to press Thomas Seebohm on several issues. Some of these—the most obvious already set into motion by the title—are quite explicit in this book, others much more implicit. I would like to close by indicating some of the questions that struck me as pointing to potentially fruitful reflections emerging out of Seebohm's work.

One question, which I hope to have at least motivated in the preceding, would be whether Seebohm saw any relation between the origin of the motivation for the generation of higher order understanding in the disruption of elementary understanding on the one hand, and the temporality of mathematical thinking that he at least implicitly suggests on the other. Could the establishment of those temporal markers at the heart of counting and measurement, making possible our bringing into view the equivalencies at the heart of ordered sequence and set (and by extension the idea of an ordered set), be understood as an at least mediated response to a need to make the world more stable, more fixed, and secure? In other words, does the role of mathematics in the lifeworld, or what Seebohm calls "lifeworlds with sciences," involve, on a fundamental level, the ordering of the disorder of time, the attempt to repair the disruptions in our lived expectations of order and continuity?

Much is made in the literature on the history of mathematics about our supposedly intuitive, natural sense for continuity, and how it in turn motivates as well as disrupts mathematical innovation (here one thinks of issues around the history of the calculus). Seebohm, it seems to me, in effect makes an interesting, provocative suggestion of something like a deep history of the human sense for continuity, one that is saturated with a poignant sense of the discontinuous, the disruptive, and related phenomena. The suggestion implied might be that the historical development of mathematics is from within anything but seamless, that it is instead characterized by contingencies, breaks, obfuscations, and the like, and that it is not in *spite* of these irregularities but *because* of them that mathematical thinking develops. Again the history of the calculus comes to mind as an example: the overall orientation since the Greeks of geometrical thinking over arithmetical speculations, in part the result of the discovery of incommensurability by the Pythagoreans, in part the success of Euclid in providing a compelling model of mathematical thinking, was an almost constant impediment to the rigorous development of the calculus right up until the breakthroughs of abstract number theory in the nineteenth century with the work of Weierstrass and Dedekind. Yet at the same time, the geometrical tradition, beginning with the development of the method of exhaustion and its creative applications in Archimedes, also served as an important incubator for not only the development of the calculus as a distinct mathematical discipline, but as a tool for mathematical physics and the study of nature generally. However predominant the prejudice of geometrical reflection, it never ceased to operate in what amounts to a flexible horizon of mathematical orientation that never ossified into a uniform, static development.[4]

Another would be the question of whether the generative phenomenology that Seebohm develops and systematizes in this book could be a fruitful point of departure for a discussion about the current state of mathematics today. Doubts about this are at least indirectly signaled by Seebohm, who seems to take it as the consensus position that Gödel's incompleteness theorems have shown that Hilbert's finitist program in mathematics is impossible, and to the extent that Husserl's theory of manifolds relies on a concept essentially equivalent to Hilbert's notion of completeness (namely the idea of a "definite manifold"), then Husserl's conception of mathematics should likewise be relegated to the dustbin of history. Seebohm, however, tries to salvage things for Husserl by making a distinction between what he sees to be the *normative* focus of Hilbert and the *descriptive* focus of Husserl: "There is no reason for a phenomenology of mathematics to adopt Hilbert's rigorous normative principle. Phenomenological accounts are descriptive, not normative." (2015, 207)

This may be a good assessment of the current state of opinion, but I am not so sure I would rest content with it. There are lots of ways to quibble: for one, there are several limited applications of Hilbert's program that have been shown to be quite successful even after Gödel's bombshell, suggesting that the concept of formal completeness established using finitist methods might have some legs yet in at least some subfields of higher mathematics and proof theory.[5] Second, the very distinction between normative and descriptive that Seebohm evokes here strikes one as problematic. Hilbert may be embracing a norm when he espouses strictly finitist methods; yet the completeness proof that would secure finitist mathematics itself is to be mathematically established, and in this sense, it represents more of a goal than a norm. That Gödel's proofs show the impossibility of such a completeness proof forces us to precisely rethink the norm adopted or, better, the basic conception of mathematics that it seeks to enshrine. This cuts the other way around, too: a philosophical clarification of the foundations of logic and mathematics for Husserl was meant to be, since the *Logical Investigations*, both a descriptive and a theoretical

[4] This role of geometry is presented in detail in Boyer 1949.

[5] One of the best recent studies of the history and horizon of Hilbert's Program is Sieg 2013.

science, but its ultimate aim was to provide the theoretical basis for logic as a properly *normative* science, not a merely descriptive one (Husserl 1975, §§4–16). Norms are blind unless inwardly determined by rigorous (theoretical) descriptions, and such descriptions are empty exercises unless they help secure the validity of norms.

A more convincing approach, it seems to me, would be to step back and observe that Gödel's proof turns on taking as paradigmatic in meta-mathematical reasoning the deployment of essentially *mechanical* methods. This was the upshot of Hilbert's finitism: the proof of consistency (or the demonstration of its impossibility) is to be established solely on the level of the formal system itself, constructed in accordance with defined rules of operation on a finite set of symbols—in short, a profile best instantiated by something like a machine. The same is true with Alan Turing's work on the decision problem, and the idea of a Turing machine best fulfills the profile of a formal system as it was conceived in Hilbert's program.

And it is precisely on this point that Seebohm's approach (as well as Husserl's, which Gödel himself believed[6]) can be instructive. Anyone who has just read Seebohm's book would immediately recognize the importance of the theme of how mathematical operations become thematized and re-thematized in ever more abstract forms in the expansion of mathematical objectivity, up to and including Husserl's theory of manifolds and modern axiomatics generally. A phenomenological reflection on the specific figure of a mechanical operation as it is articulated in modern formalism, a reflection that seeks to articulate its sources and limits as a species of mathematical cognition, strikes me as a potentially fruitful way to think through the meaning of Gödel's incompleteness theorems for our understanding of mathematics generally. For it opens the way to making a potential distinction between mechanical and non-mechanical, but equally logically rigorous forms of mathematical thinking, or more generally, between the kinds of objectivities that are instantiated in formal deductive systems, and those that fall outside of their specific purview of definability. It offers, in other words, a potential way to regain a broader conception of *mathesis* that would transcend the figure of a mechanical operation to once again include a deeper sense of the *learnable*, and without sacrificing the principle of rigor.

A more general point can be added here. The phenomenological account of the genesis of mathematical intuitivity sketched above, thanks to which primitive mathematical objects come into view by way of configurations and accomplishments that draw directly from the inner richness of our consciousness of time, can be (and to some extent in Seebohm's book is) extended to include other dimensions of the lifeworld, so, for example, the constitutive role of *traditions*. This is at the root of any phenomenological account of the meaning of "operations" with mathematical objects: such operations are constituted not only in the context of the complex temporality of mathematical thinking, but also in an environment established by a dimension of specifically *historical* complexity. Not only the techniques of mathe-

[6] See Tieszen 2011, who provides an interesting and detailed elaboration of Gödel's Platonism on phenomenological grounds.

matical thought, but the very posture and orientation of mathematical thinking generally, relies on the inner movement of an intentional history, one constituted in the manners in which the techniques of thinking are secured and transmitted through generations, or in other words in traditions. This of course takes us again to the itinerary of Husserl's "Origin of Geometry," and bears directly on the set of problems that belong to the second stage of Seebohm's analysis of the origin of science as a figure of higher understanding: namely, the emergence of idealizations. This in turn bears on another obvious problem in any phenomenological account of mathematical thinking, above all one that would emphasize the role of time-consciousness: how from within the flow of the experience of time something like ideal objectivity—which is, of course, *a-temporal*—comes to be apprehended.

It seems to me that such an approach, aspects of which are already in play in *History as a Science*, would represent a reflection that would, in ways perfectly commensurable with Thomas Seebohm's thought, open up onto a depth investigation of the historical interweaving of *mathesis* and lifeworld that would be of interest to the philosophy of mathematics generally, and the specific problems faced by modern mathematics today.

References

Boyer, Carl. 1949. *The history of the calculus and its conceptual development*. New York: Dover.

Hartimo, Mirja, ed. 2010. *Phenomenology and mathematics*. Dordrecht: Springer.

Husserl, Edmund. 1954. Vom Ursprung der Geometrie. Beilage III. In *Die Krisis der europäischen Wissenschaften und transzendentale Phänomenologie*, Husserliana VI, ed. Walter Biemel. The Hague: Martinus Nijhoff.

———. 1975. *Logische Untersuchungen. Erster Band: Prolegomena zur reinen Logik*, Husserliana XVIII, ed. Elmar Holenstein The Hague: Nijhoff.

Jones, Mark W. 2000. Doric measure and architectural design 1: The evidence of the relief from salamis. *American Journal of Archaeology* 104 (1): 73–93.

Kant, Immanuel. 1902. *Kritik der reinen Vernunft. Kants gesammelte Schriften*. Berlin: Preußische Akademie der Wissenschaften.

Lohmar, Dieter. 1989. *Phänomenologie der Mathematik*. Dordrecht: Kluwer.

Seebohm, Thomas M. 2015. *History as a science and the system of the sciences: Phenomenological investigations*. New York: Springer.

Sieg, Wilfried. 2013. *Hilbert's programs and beyond*. Oxford: Oxford University Press.

Tieszen, Richard. 1989. *Mathematical intuition: Phenomenology and mathematical knowledge*. Dordrecht: Kluwer.

———. 2011. *After Gödel*. Oxford: Oxford University Press.

Weber, Heinrich. 1891/1892. Leopold Kronecker. In *Jahresbericht der deutschen Mathematiker-Vereinigung*, 2.

James Dodd (Ph.D., Boston University, 1996) is Associate Professor and Chair of the Department of Philosophy, New School for Social Research. His research interests include classical phenomenology, the philosophy of architecture, and violence as a philosophical problem. He is the author of *Idealism and Corporeity* (Kluwer, 1997), *Crisis and Reflection: An Essay on Husserl's Crisis* (Kluwer 2004), and *Violence and Phenomenology* (Routledge, 2009).

Chapter 9
From the Epistemology of Physics to the Phenomenology of Nature: Some Reflections in the Wake of Seebohm's Theses

Pedro M. S. Alves

Abstract I discuss Seebohm's fundamental tenet regarding the scope and form of a phenomenological epistemology of natural sciences, particularly physics. I stress, first of all, that the issue is deeper than a simple epistemological approach, taking modern physics simply as a fact. I point out that contemporary physics, particularly quantum mechanics, does not deliver a unified, coherent, and uncontroversial view of Nature. Taking this into account, I argue than the task will be to return to the project of a philosophy of nature, by means of a phenomenological critique of the sense institutions that, in modern times, opened the field of physics and of a mathematical approach to nature. I claim that these original institutions are coming to a point of rupture with quantum physics, given the trend to absorb nature in the mathematical apparatus that is put forth to account for it. At the end, I stress the possibility within the contemporary debates about quantum physics of a realist account of quantum entities that restores the classical sense of nature as a morphological realm that is suited for linear approaches by means of exact essences, but which is not reducible to them.

9.1 A Way Out of the Paralyzing Dilemma of Phenomenology

Regarding the epistemological and ontological discussions of the foundations of physics, can phenomenology help?[1] Does phenomenology have something really important to say in the debate about physical reality? What sound contribution can

[1] This article is an attempt to appraise and develop the theses put forward by Thomas Seebohm in his last, stimulating book titled *History as a Science and the System of the Sciences*. Particularly,

P. M. S. Alves (✉)
Department of Philosophy, School of Arts and Humanities, Centre of Philosophy, University of Lisbon, Lisbon, Portugal

© Springer Nature Switzerland AG 2020

T. Nenon (ed.), *Thomas Seebohm on the Foundations of the Sciences*, Contributions to Phenomenology 105,
https://doi.org/10.1007/978-3-030-23661-8_9

a phenomenological epistemology and a phenomenological philosophy of Nature give to these issues?

As Seebohm writes, "The first task for a phenomenological epistemology is … to ascertain whether, where, and how reflective phenomenological descriptions can be of significance for the epistemology of the natural sciences" (2015, 162). In fact, as he stresses at the very outset, positivism, and later analytic philosophy, "governed the epistemology of the natural sciences in the nineteenth and the first half of the twentieth century" (161). They were fully wed to a reductionist approach to the system of the sciences, that presents physics as the last and most fundamental science, and to a phenomenalism without any further ontological commitments. It is to be expected that a phenomenological epistemology and phenomenology of the region of nature will defy both of these assumptions. However, in the absence of a long-established tradition of a phenomenological epistemology of the natural sciences, phenomenology must, so to speak, begin by proving its own relevance for an epistemology of the natural sciences, and finally for a philosophy of Nature.

Still, in my opinion, the task is even more profound because there are internal trends in phenomenology itself that prevent it from having an immediate and straightforward dialogue with natural sciences. This is so because, regarding these matters, phenomenology was (and still is) caught in a dilemma. To put it simply and directly, what I call the "phenomenological dilemma" regarding Nature and natural science, particularly contemporary, non-classical physics, amounts to this: what science tells us about Nature *as it is in itself* is largely beyond the conceptual reach of phenomenology, and what phenomenology can say about the *sense of the givenness* of Nature is almost irrelevant for natural science. As a result, phenomenology faces the dilemma of being silent about Nature due to the science of nature, or of developing a philosophy of nature by turning its back on, and even criticizing, the contemporary science of Nature.

Certainly, things are not so black-and-white. One must do justice to those many scholars, from Weyl and (the earlier) Becker to the present day, who have dealt with issues regarding the foundations of physics or mathematics from a phenomenological perspective. Nevertheless, as a reflective analysis of sense formations, phenomenology is much more at ease with problems related to the social world and to the intricacies of the conscious self. In addition, regarding the epistemology of the natural sciences and the philosophy of Nature, a tendency (which I deplore) arose early on and strengthened itself afterwards in the phenomenological tradition. I am referring to the claim that, in all these matters, phenomenology will be the carrier of, and, thus, will give expression to a more fundamental point of view regarding Nature itself. This point of view had supposedly to do with an original *experience* of Nature, as against Nature in the "garb" of the conceptual and mathematical rationality that was superimposed on it. This latter clothing would be a loss of the original meaning

this article deals with chapters 7 and 8 of Seebohm's book (pages 141–254), which are a discussion of the generative origins of modern natural sciences, especially quantum physics.

of Nature, and thus a departure which would have had dramatic consequences regarding thinking itself and the very *ethos* of humanity as such.

Surely, these trends are more apparent in Heidegger's sharp criticism of scientific rationality, as well as in a particular reading of Merleau-Ponty's thesis about the "primacy" of perception. Nevertheless, some important aspects of Husserl's thinking are ambiguous enough to suggest this kind of interpretative direction as well as its opposite. I have in mind chiefly these:

(1) The emphasis on the lifeworld as the primitive form of world experience.
(2) The determination of the object of physics as a logical and theoretical "substruction" [*Substruktion*] put under the things of the perceptive world.

As a matter of fact, one can stubbornly remain in the lifeworld, affirming something like this: there is only one true world, our lifeworld, and, in comparison, the theoretical world of natural science is no more than a farewell that brings us nowhere; the world of physics is a world where we cannot live, and then a departing from the original place (and theme) of all human thinking and life. According to this view, its unique efficiency and reality would be the technical transformation of the lifeworld, along with the emergence of a human behavior increasingly mediated by instruments (the *homo technicus* and *instrumentalis*). This way of thinking goes hand-in-hand with an instrumentalist account of science and with an ontological quietism about its objects. Husserl himself is often viewed as someone who espouses such kinds of views, namely when he emphasizes that the world of science is an alienation from the lifeworld, and when he puts in that alienation the very roots of the crisis of the European sciences.[2] However, an *Entfremdung* is certainly a departure; but a departure from the lifeworld can also be a *discovery* of a new world, so that one can also go in the opposite interpretative direction, trying to account for the complex net of meaning-institutions that both connect and separate the lifeworld from the world of physics. There are continuities and ruptures of sense that must be accounted for. In my opinion, the point for Husserl was that this genetic process was wholly concealed through the substitution of original intuitive thinking by mechanical methodological procedures. As he said in one of his acute statements, immersed in methodological techniques forgetful of the original institutions of meaning underlying them, physicists would have become simple "engineers of science" that, as thinkers, would have fallen in the "stage of naiveté" (Husserl 2002, 9–10).[3] Now, recovering these genetic processes is tantamount to submitting to a phenomenological descriptive analysis (1) the theoretical attitude of the modern natural sciences, and (2) the constitution of their thematic objects. Opportunely, this is precisely the path taken by Seebohm with his inquiry concerning the "generative foundations" of

---

[2] The interpretation of Husserl's philosophy of natural sciences as a kind of positivism and instrumentalism was recently submitted to a severe criticism (convincingly, I believe) by Lee Hardy (see Hardy 2013).

[3] "Die Naturforscher wuchern mit einem ererbten philosophischen Pfund gewissermaßen als Techniker, sie sind gleichsam zu Ingenieuren der Wissenschaft geworden … Aber als Denker sind sie auf das Stadium der Naivität zurückgefallen."

natural science within the lifeworld, along with his description of the two "abstractions" that circumscribe the object of the natural sciences ("soft" and "hard").

Furthermore, it is certainly true that Husserl characterizes the positing of a "true world" as correlate of the exact sciences of Nature as a "substruction." This has three dimensions. On the one hand, the world "sub-posited" as true remains inaccessible for perception, as something that, by definition, is not related to an original experience of its proper being. On the other hand, it is posited precisely as *true*, i.e., as the objective world against the subjective-relative significance of the lifeworld. Finally, this "sub-structure" [*Unterbau*] is the correlate of a "logical-theoretical" intentionality, where the modes of access belonging to the noematic sense are almost shaped by mathematical formalism, depending on the abstraction processes that lead to exact and formal essences.

As I see it, pointing to these features is not a critique of the achievements of natural science. It is rather the outline of an entire phenomenological research program, thematically related to the "phenomenon-natural-science." In my view, the word "logical" must be interpreted in the wider sense of the *mathesis* as a theory of multiplicities, and the "theoretical" means here the intentionality that constitutes the sense of a being *in itself,* beyond any subjective-relative mode of appearance in the perceptual world. It involves a conceptual determination of what is *real,* and the definition of the experimental procedures in which it becomes *observable*. These experimental procedures refer back to a mathematical framework, which defines the quantifiable aspects of the physical reality that is posited. They determine the modes of access to that physical reality. For this very reason, this "being in-itself" is precisely the correlate of such complex constitutive processes, and not a thing that will be neither accessible nor observable in the Kantian sense of a *Ding an sich.*

9.2 What Calls for the Philosopher in Contemporary Physics?

It is, then, a great merit of Seebohm's book to make phenomenology cope once more with the epistemological problems of the so-called "hard sciences," especially the physics of the last century. Indeed, in two sound, well-informed, and insightful chapters, Seebohm addresses fundamental issues regarding the epistemology of natural science from a phenomenological vantage point, and especially the epistemology of physics.

Seebohm rightly stresses the difference between classical and post-classical physics based on a mutation of their mathematical formalisms (2015, 222). The mathematical framework of classical physics was based on exact essences that were an idealization of things or phenomena previously given in the lifeworld (like mass, acceleration, and so on). By contrast, post-classical physics makes use of formal essences entrenched in mathematical formalisms that are very far from any pictorial representation. In addition, while great developments in mathematics were often prompted by physical problems (for instance, the discovery of calculus by Newton), the formal mathematics used in post-classical physics represents only a part of the objects created by pure mathematical thought. This is certainly true for relativity, which makes use of non-Euclidean geometries developed for the first time by Riemann as a pure formal theory of manifolds. From this point of view, relativity and quantum physics share the same position as non-classical physical theories.

However, in my opinion, what was truly new in the science of Nature in the last century was put forth by quantum physics. It is no longer a "classical" theory, even though relativity continues to be to a certain extent "classical," and can be viewed as the culminating point of mechanics (reframing Newton's gravitation law) and electromagnetism (without the ether-hypothesis). Certainly, the four-vector geometry of special relativity overthrows our elementary understanding of the nature of space and time, and the same happens with the metric tensor of general relativity. Nonetheless, the whole theory follows directly from two basic (and comprehensible) principles whose physical self-evidence is independent of the mathematical formalisms to which they give rise. First, the principle of the invariance of the speed of light for all frames of reference, which is unavoidable when the ether is abandoned, and, secondly, the principle of general co-variance, which states that the laws of physics shall take the same mathematical form in all frames of reference. Certainly, the huge modifications of space and time, put together in space-time metrics, are not accessible without the mathematical formalism in which they are formulated. "Space-time," which is already an interpretation of the physical meaning of the four-vector in the light of our *experience* of Nature, is in itself a "mathematical-physical" object, as Seebohm points out, i.e., a physical reality that is only accessible by means of the mathematical formalism that determines it. However, relativity has a clear logical framework and a well-defined underlying ontology.

Regarding quantum physics, the problems a philosopher must address are far more demanding than in the past, including the case of relativity. In fact, classical mechanics, relativity or electromagnetism conveyed a clear and unambiguous ontology, respectively centered in the concepts of mass, distributed over space and time, of field, as an extended, non-punctual reality, and of space-time "curvature," in its interactions with the stress-energy tensor. The mathematical formalisms they developed were closely related to a clear conception (and even to an image) of what the physical reality was in and by itself.

The other way around, quantum mechanics developed a mathematical formalism which was largely *undecided* about the very nature of the entities to which it referred. This problem plagued quantum mechanics since its very beginnings (with Bohr, Louis de Broglie, Schrödinger, Born, among others), and still continues today. In addition, there are many other difficult aspects of quantum physics regarding both the *depiction* of the physical reality and *what* should be accounted for as "physical." Thus, concerning quantum mechanics, we are not in a somewhat Kantian situation. We do not have a full-fledged, uncontroversial "*fact* of science" with its fixed ontology, so that the task of the philosopher would simply be to proceed back from this fact to its "transcendental conditions of possibility." In its place, we have an accurate mathematical formalism (perhaps, the most accurate science has ever created) and a *problem* regarding its ontological interpretation in order to characterize what is "physical." This is the reason why the task of the philosopher here will not be as "simple" as it was in the past. The question is not only to reflect upon the fact of science as an epistemologist; instead, the task amounts to entering into the controversy about what physical reality *is* in itself, not merely as an epistemologist, but rather as a philosopher of Nature. The modern philosophy of Nature was written in mathematical formulae; now, with quantum physics, the mathematical formalism only promises a philosophy of Nature. The task of the philosopher begins right here.

Regarding the demanding aspects of quantum physics, I offer only an outline.

With quantum mechanics, we have:

1. *A non-classic interpretation of probability.* Unlike Boltzmann's probabilities, quantum probabilities do not reflect our ignorance about the individual real elements of a complex system. They represent the likelihood the interaction with a measuring device will *create* a certain outcome *as the actual state* of a physical system. It was Max Born who proposed this probabilistic interpretation of Schrödinger's wave-function. The square of the amplitude of the wave-function in same region of configuration space is related to the probability of finding the particle in that region when and only when a measurement is made. So the wave-function does not depict a quantum system, but only our knowledge about the probability of certain later outcomes. Projecting this onto an individual system entails that it will be a superposition of *all* its *possible* actual states. In fact, according to Dirac, quantum mechanics stresses that whenever a quantum system can be in a multiplicity of states, the superposition of states is itself a state in which the system is until some measurements are made.

The baffling consequence is that, *before measurement*, quantum reality evades itself into mathematical virtuality. A quantum system is not determined in itself. In addition, there are non-commutative operators (like position and momentum), so that measuring one of them implies a disturbance in the other (the lesson is that measuring first A and then B is different from measuring first B and then A). As the "normal" interpretation of quantum mechanics claims that it is *complete* (as far as I know, no other theory has claimed to be a complete depiction of a physical reality), with no place for "hidden variables," one must conclude that there is no *objective, determinate* situation of a physical system before measurement. Prior to that, there is only the superposition of all possible states that the mathematical formalism describes. This is tantamount to concluding that the *physical* and the *mathematical* are here fused in such a way that the classic idea of *reality* is destroyed, i.e., the idea of something that is in *itself* completely determined, and not a mixture of mutually incompatible states.

2. *An appeal to consciousness in the passage from virtuality to actuality.* The mathematical framework of quantum mechanics does not describe the passage from potentiality to actuality. The so-called collapse of the wave-function is simply supposed to occur, so that the system can pass from the superposition of all its

possible states to an actual definite state. Facing the infinite regress implied in the fact that the measuring apparatus is itself a quantum entity, and that even the sense organs and the brain are quantum systems entering in a superposition with the physical system under observation, von Neumann introduced his famous thesis in which the wave-function collapses when (and *if*) the system interacts with consciousness. Only consciousness is supposed to be a reality beyond the physical realm that puts an end to the infinite regress. However, not a word is said about the way a supposedly non-physical entity can interact with a physical entity and prompt the passage from the superposition of states to one and only one definite state.

As a consequence, physics, the traditional bulwark of realism, is now strangely suggesting a somewhat anti-realist, even idealist, mind-dependent conception of reality.

3. *Entangled states and non-locality.* Two particles that have interacted are described by a single two-particle state vector. When one is measured, the state vector collapses into a definite state, forcing the other particle to assume instantaneously a specific correlated state, no matter how far it will be from the former. Bohr abandoned the idea of a "disturbance" introduced by measuring operations in his answer to the EPR argument in 1935. Since then, he defended an indeterminacy of the system before its relationship with an apparatus, which is tantamount to a real action at a distance in the case of entangled particles. This was a direct vindication of non-locality. The particles lost their individuality in space and time. Bell's inequality theorem and experiments promoted by Aspect and Wheeler gave some experimental, albeit very controversial support to the reality of entanglement.

As a consequence, instead of referring to a "spooky action at a distance" or to an instantaneous transmission of signals, one must ponder the very adequacy of the concepts of space and time to account for the quantum world. Here, the "proximity" and "distance" between two particles are not measured by metrical space, but by the directness of the interactions they have with one another. Two particles can be very far from one another at the scale in which we perform spatial measurements, while, at the quantum scale, they are non-separated and each one is, so to speak, "on top" of each other.

4. Last but not least, *an unresolved corpuscle-wave dualism.* Indeed, since its very beginnings, an undecided corpuscle-wave duality was entrenched in the very heart of quantum mechanics. Classical physics made use of two divergent idealizations: the wave and the tiny, at limit punctual corpuscle. Both have their roots in the immediate experience of Nature in the lifeworld. The controversy about the nature of light, from Newton and Huygens onwards, was based on the incompatibility of these two models. Now, the modern double-slit experiment (the original was made by Young with light in 1801) showed that quantum entities should be accounted for *both* as corpuscles and as waves, depending on the kind of experimental apparatus that was used (one slit open or two slits open). This

situation is not specific to light. The same results can be obtained with electrons, protons, and so on. As a consequence, matter itself displays properties that can be accounted for as classically defined waves *and* as corpuscles. The wave-like character of the so-called "matter" was famously proposed by de Broglie (1924). Regarding this inexplicable dualism, Bohr put forth his principle of complementarity (Bohr 1928). The wave-like character of "corpuscles" or the corpuscle-like character of "waves" (in this case, converted into mere mathematical entities, as in Born's interpretation of the solution of Schrödinger's equation [Born 1926]) remains an unresolved controversy at the most fundamental level of quantum physics. Heisenberg and Born took the side of the corpuscular depiction, while de Broglie, Schrödinger and Bohm took the opposite side of a wave-like interpretation. Both interpretations of the nature of quantum entities fit in the experimental evidences and are mathematically equivalent.

As a result, it seems that no *internal* criteria are available (I will briefly return to this point later on). In such a case, we could do away with the entire question. However, this is a paramount issue one cannot simply ignore, as it concerns the very ultimate nature of physical reality. The question concerns *what type of external criteria* are available and are justifiable in order to develop an ontological interpretation of a "silent" mathematical formalism. One can suggest that the problem is, after all, indifferent for a phenomenological epistemology. Instead, I will suggest that this is a real problem for a philosophy of Nature, whose criteria should be searched for in a phenomenological enquiry into the regional ontology of Nature, and that, as logic teaches, when two universal statements contradict each other ("All quantum entities are corpuscles", "No…"), they can both be false.

All in all, the classical conception of *reality*, and the traditional way the *physical* was understood as different from the pure *mathematical objects* used in calculations and previsions was put into question to an extent we can only guess. This is not a lesson of quantum physics itself, but of a particular (and peculiar) construal of the mathematical formalism called the "Copenhagen interpretation." It was roughly put forth by professional physicists like Bohr, Heisenberg, Born, or von Neumann. They were, simultaneously, the brilliant creators of quantum physics, and its interpreters as philosophers of Nature. The prestige they acquired as physicists was projected into their interpretative work. Nonetheless, the interpretation they proposed of *physical reality* strikes the professional philosopher as ontologically incoherent and plagued with illogical statements. We could say: if quantum mechanics contradicts classical logic, so much worse for logic—instead, we must develop a "quantum logic." Seebohm believes that this is an *ad hoc* response that begs the question.[4] As do I, with an additional reason: the formalism of quantum physics relies on linear algebra, which is tributary of the traditional bivalent logic. The logical problems arrive only at the level of a particular, and controversial, interpretation of the formal-

[4] "Quantum logic is, hence, nothing more and nothing less than a re-formulation of the ontological problems in the level of a formalized modal logic, and not a solution for the problems" (Seebohm 2015, 227).

ism. In addition, (1) the actual superposition of mutually incompatible states, (2) the wave-corpuscle dualism, (3) observation as somewhat creative of the very reality observed, and (4) the completeness of the quantum description, as if there was nothing more in the quantum world—all this strikes not only the realism and empiricism of the scientific stance, but claims for an ontological clarification of the concepts we use in order to understand the most fundamental dimension of reality: *phýsis* or Nature.

So, I will ask:

What has *phýsis* become in light of quantum physics?
What was *phýsis* for the physics of modernity?
How do these accounts of *phýsis* relate to the *sense* of Nature that opens the field in which, afterwards, the physicist enters as a methodical researcher?

Albeit without good answers, these are my questions.

9.3 The Thematic Attitude of the Modern Physics: Causes and Consequences of the "Second Abstraction"

Everywhere we find the idea that the perceptual world is simply a manifestation (for us humans) of an underlying reality, and that this world that perceptually appears is driven by forces that are only accessible in a particular stance. This is the case for religion and mystical stances, which purportedly convey knowledge of (and a control over) that hidden reality. The same happens in pre-scientific theoretical thinking (as a contemplative stance, not appointed to the control of Nature), which was developed in the framework of the metaphysical and theological conceptions about the ultimate nature of reality.

The specific historical situation of the lifeworld of modernity is that it is a lifeworld both with sciences and with several philosophies of Nature. It is within this particular historical context that the modern positing of a reality that is behind perception and would be accessible "as it is in itself" by means of some reframing of the perceptual givenness develops. For the Moderns, the "reality" that manifests itself in perceptual acquaintance is the *natural* or *physical* world. The leading question is not how the perceptive objects are given—this is a matter of a phenomenological description of perceptual evidence. The question is neither to collect facts and regularities that are already accessible in experience through the cognitive methods of elementary understanding—this is a question of a Baconian conception of science. The questions that move modern science is why this world is there with such and such things, and what is the underlying composition of the objects given in perception—this is a generative question, and the answer goes beyond the mere givenness of the perceptual world right to its underlying "laws." "Beyond" or rather "under" perception is the modern mode of access to the *physical.* This conveys the insight that the fundamental dimension of reality (which a reframing of the perceptual acquaintance would attain) is not mind-like, and much less spiritual in a religious or magical sense (as it still was for certain conceptions from the Renaissance). Instead, it is a realm we can only access if we methodically abstract from every aspect of our human take on the perceptive world. This has to do with what Seebohm calls the "first abstraction" constitutive of the thematic object of the natural sciences of modernity (both "soft" and "hard"). In a phenomenological framework, it amounts to a *methodological exclusion* regarding the objects in the lifeworld of the properties that refer back to social, cultural, or psychological sources of meaning. The given is thematically "limited" to the dimensions that sheer perceptual acquaintance can exhibit—corporality, duration, extension, along with the so-called "secondary properties." In addition, as Seebohm also emphasizes regarding the modern conceptions of Nature, we find as a historical but decisive contingency (1) the downfall of Aristotelian ontology, and (2) the recovering of the ancient atomistic ideas. The methodical abstraction from purposes, final causes, intentions in general, intelligent design, as well as the criticism of the Aristotelian concept of formal cause spring from this new conception, which opposes the *physical* to the *psychic*. This is the reason for the denunciation of scholastic physics as wholly anthropomorphic (Descartes is here a reference point). In addition, the general conception of the physical as constituted by small corpuscles retrieves the ancient atomistic ideas. This general conception of the "external world" (i.e., alien to the "mind") can be found in Descartes, Boyle, Locke, Gassendi, Hobbes, and even in Newton. The first abstraction and the split between properties of matter and properties of the mind are parallel and reinforce one another. End-directed processes belong exclusively to the "mind" or to an illusory "mentally-laden" apperception of Nature. (2015, 170). It is a reduction that operates within the first abstraction. It gives up the "secondary qualities" as being merely subjective and delivers a conception of reality according to which what is "physically real" is *nothing more* than what is quantifiable and expressible in some mathematical formulae.

Measuring for the practical ends of life is a generative foundation of this modern conception of the physical, as Seebohm points out. However, I believe that it is not a sufficient condition. The question is no longer that we measure physical objects. The question is that here the physical *is only what can be* measured. Where did this new insight come from? I retort: from the heavens, so to speak. In fact, long before modern natural science appeared with Galileo and Descartes, there was a science entirely devoted to measuring motion in space and calculating the trajectories of bodies. It was quite successful in joining mathematical models and observational data. I am referring to astronomy, which developed as a mathematical science since Ptolemy. The Aristotelian divide between sub-lunar and supra-lunar worlds prevented the building of an earthly physics in the same way. In addition, there were many of phenomena that could not be accounted for as a result of simple local motion: generation and corruption, growing, becoming, and especially patterns of *organization,* which the Aristotelian ontology captured under the concept of *form.* However, Copernicus showed that Earth was a *celestial* body like others: it revolves around the sun like the other planets. For me, this is a highly important result of the so-called "Copernican revolution." For the development of the new physics, it is even more important than the overthrowing of the "central position" humanity had in the former cosmic order. Indeed, the homogenization between Earth and the planets that this new vision entails will have two important consequences. On the one hand, as the celestial bodies had the same nature as Earth, this move opened the path to a celestial *mechanics*, which came to its ultimate form in Newton's law of gravitation. This is well-known. On the other hand, there is an aspect which is seldom referred to: this move also suggested the possibility of an *earthly* physics entirely based on the phenomenon of local motion, developing the suitable mathematical formalism. These were the great insights of Galileo. Firstly, his observations of Earth's moon and of Jupiter's satellites carried evidence that the planets were of the same physical nature as Earth. Secondly, the mathematical laws of the oscillation of a pendulum, of the falling bodies, the parabolic trajectories of projectiles, the principle of inertia, and the Galileo-transformation for the addition of movements, as it is called today, all this showed the possibility of a mathematical physics entirely centered on the phenomenon of local motion. With Descartes, albeit his refusal of the void, the corpuscular hypothesis furnished the required image of a physical realm where the basic phenomena were cinematic, dynamical, and mechanical. Hence, the modern concept of "matter" was a complete reversal of the Aristotelian one. Matter was no longer an undetermined substrate; on the contrary, matter was an autonomous realm which had its own laws: the laws of motion and communication of motion.

What consequences followed concerning physics? Here, Descartes is the leading theoretician of the epistemological and ontological foundations of modern physics. Let us take a closer look at what Descartes's point really was. It was a very bold bet: nothing less than systematically suppressing in the description of Nature the reference to patterns of organization—which were entrenched in the concept of "form"—in order to come to them as results of the mechanistic, blind, and not teleological local motion of matter alone. That is to say: Aristotelian-scholastic physics took a top-down perspective. It started with the organization-pattern, called "form," and proceeded to the material base in which it was stamped, continuing successively on until the ultimate "undetermined matter," the so-called *substratum*, was reached (only in thought, for sure, because there is no indeterminate matter). Nature was seen as a system of forms. By contrast, Descartes's bet was to proceed from a bottom-up perspective. He began with autonomous laws of the motion of matter in order to arrive at the organized structures not as principles, but as results of these core laws. Given that the reference to "form" (i.e., to organizational patterns) was removed from the study of Nature, physics had to discuss the relationship between such scalar and vectorial quantities as force, movement, velocity, acceleration, mass, and so on. Regarding organized top-structures, like living organisms, old names such as "substantial form" and "soul" were thus no longer applicable; considering that they were explainable in terms of the local motions and arrangements of their constitutive parts, their new and true name was instead (natural) "machines."

Surely, the old concept of "form" was poorly elucidative, as it really did not explain anything. Notwithstanding, it had several advantages in the understanding of Nature. Namely, it directed the attention from the very start to patterns of organization, and not simply to bits of bare mass" (the "*quantitas materiae*"), and, in addition to this richer point of view, it endorsed a broader concept of movement, which encompassed not only local displacement, but also the phenomena of growing and alteration that were characteristics of structured and organized beings. The Cartesian "reform" of physics was, then, based on several very strong assumptions. Firstly, that bare matter could be considered by itself as a closed realm of local motion and forces acting by impact, that is, that kinematics and dynamics were the very core of physics; secondly, that organized natural structures could be explained mechanically, without the involvement of teleological principles and final causes; thirdly, that every time we were talking about something like a "plan," an "organizing principle," or an "intention," we would be talking about a mind, i.e., about a psychic being and definitively not about physics.

What we gained by this *tour de force* is well known to all. There is no need to describe it again. Modern natural philosophy took physics as the fundamental science, and physics was defined as the science of local motion until Faraday and Maxwell's works (in the nineteenth century) on electromagnetism introduced a new branch. In line with this leading idea, Kant tried to present nothing less than an a priori deduction of the object of Physics. In his *First Metaphysical Principles of the Science of Nature,* he presented physics as the science of matter, and the science of matter as the science of motion, to conclude, supposedly still in an a priori fashion, that the several parts of the science of matter in motion were phoronomy (kinematics), dynamics, mechanics, and phenomenology (in a very particular sense).

However, Kant himself realized at the same time the limits of mechanistic explanatory schemes, especially to account for the phenomenon of life. What he wrote in the third critique was another expression of the thesis that organisms would never be explainable in terms of mechanics. Let us return briefly to Descartes's theoretical decisions. They were certainly a tremendous step forward, but they were also a dramatic impoverishment. First of all, there are natural processes that seem to be driven by something like a plan, or a pattern, that determines in advance the disposition and the arrangement of the parts. Living organisms are an example of this. But global arrangements like the ones we can see in crystals, in snow or ice patterns also suggest to the natural philosopher that there is something like a "natural technique" or a "natural plan," that is, something like an unconscious natural intention (or propensity) in several physical processes. The core of this insight into the deep structure of physical beings lies in organization as such, not in finalistic or intentional processes, which are just an interpretation about *how* organization appears. Indeed, final causes or the reference to a "natural technique" or a "natural intention" are a rather bewildering response to the problem of organization, i.e., to organization as a phenomenon of nature in need of a *natural* explanation. We can now see what happened: Descartes's reform suppressed from the science of nature the reference to final causes or intentions acting in natural phenomena. In doing so, organization as a primary fact (to which the concepts of final causes and forms used to apply) vanished from the regard of the natural philosopher. Refusing a bad answer to the problem of natural organization patterns (substantial forms, final causes, and so on), Descartes imposed that the very fact of organization be also dropped; better, what was dropped were the concepts that constitute organization *as a fact*. This is precisely the first impoverishment I want to emphasize: organization was formerly picked up in nature through the concepts of forms, inner intentions, ends, and plans; all those concepts were now interpreted through the model of the purposive action of a conscious mind, and, for that very reason, they were excluded from the science of matter; as a result, organization was regarded only as a secondary upper-level phenomenon of something more fundamental, namely, as a "blind" consequence of the pure and simple local motion of matter. Now we can see the second dramatic impoverishment Descartes's reform brought with it. Organization is a high-level phenomenon that is not explainable by a linear reasoning ranging from the parts to the whole and considering the whole as the simple sum of its parts. Organization is a pattern that emerges on an underlying multiplicity of elements, but in such a way that it *seems* that it was the pattern that directed in advance the disposition and arrangement of these elements. This was the "intentional" or "teleological" delusion to which the concepts of substantial form and final causes gave expression to. In refusing teleology by assimilating all forms of design to the purposive action of conscious minds, Descartes's reform closed at the same time all direct paths to the organization-phenomenon. Thus, the only available intellectual scheme for explaining the top properties of an organized natural structure was to try to recover them as an effect of the blind arrangement of its parts, as if they happened by pure chance.

The problem is that organization as a natural phenomenon is *self*-organization, and self-organization cannot be accounted for with such a poor concept of matter. The first difficulty is that this matter does not contain dynamical power by itself (which originated a huge controversy in modernity). The second problem is that a self-organized system is not explainable as a whole, which is analyzable in parts and reducible to the properties of its elements.

We can suppose that this issue starts only at the level of organisms, i.e., living bodies or the "lowest level of second-order wholes" (Seebohm 2015, 242). Still, my guess is that organization begins much sooner in Nature: *it is Nature itself.* So, I ask: can we take the concept of a self-organized whole as the very definition of a natural being?

I will briefly argue this point. It will give me an argument for a reappraisal of the quantum description of Nature in light of the so-called Copenhagen-interpretation.

9.4 The Sense of Nature and Physics

The perceptual acquaintance with objects in the lifeworld has been a long-standing theme of phenomenological analysis. By contrast, the categorization of the objects of perception is, in my opinion, an issue that deserves a more detailed analysis. Considering the region-Nature, Husserl's approach in *Ideas* is in line with the modern way of thinking Nature as a "material" realm. He distinguishes three strata of sense: the *res extensa,* the *res temporalis,* and the *res materialis* (1977, 347–348).

This simply amounts to repeating the modern characterization of Nature: everything is a unity of duration (in chronological time), it has a form (*Gestalt,* i.e., a place in space, not a *morphé*), and finally it is a "substantial-causal" thing in a net of interconnections with other things. Space and time are dependent moments of the *res materialis.* They can be treated in an aprioristic way by means of exact essences (in chronometry or Euclidean geometry) or formal essences (in the framework of a theory of manifolds). Regarding the *res materialist* itself, it is characterized by its causal dependencies, which are a matter of empirical research and of mathematical formulation by means of idealized objects such as mass-points, vector or scalar fields, forces, etc.

However, in my opinion, this analysis misses an important point: what are the conditions for something to be perceived as a *natural unity*, i.e., precisely as a *thing* in the strong sense of a *concretum?* Modern physics espouses an atomistic conception about its objects. By "atomism" I do not refer to a particular historical doctrine, but to an approach which intends to analyze complex systems in order to find its supposed ultimate elements, and, based on them, to reconstruct the former wholes by adding "forces" of cohesion between those now separated elements, as if they were the building blocks of reality. In this sense, the concrete, natural unities will be the non-analyzable ultimate elements we will eventually reach when we reduce biology to biochemistry, then to chemistry, then to atomic physics, and finally to the sub-atomic realm.

The reductionist program of positivism is entirely in keeping with this way of thinking. Nevertheless, I think that there are two important arguments against it.

Firstly, a scale problem: when we reduce a realm to another supposedly more fundamental, we lose sight of the specific organizational laws of the reduced realm. For instance, we cannot rewrite biological laws in terms of chemical laws, even less in terms of atomic laws. On the contrary, we must say that biological phenomena are correspondent with some lower chemical phenomena. Nevertheless, the biological realm adds specific constraints to the chemical, so that some chemical processes are enforced, while others are inhibited. Take for instance the chemical constituents of a cell. As independent elements, they can react with one another in a multiplicity of ways. However, when entrenched in the cell, only a part of the set of all possibilities of chemical reactions is allowed. These constrains came from the top-realm and *define* it, so that they cannot be recovered at the lower-level realm.

Secondly, as against the atomistic-insight, we must stress that a *natural unity* or *concretum* is always apprehended in the perceptive lifeworld as a *whole* whose parts stand in a particular relation, and not as a non-analyzable unity. Certainly there are many wholes that, at least ideally, can be analyzed into ultimate pieces according to a relation of foundation. Because of this, they are perceived as being already *aggregates* and not as natural, individual beings in the pregnant sense. This is a lesson we can read in the perceptual acquaintance. We perceive something as a natural unity when, going from simple seizing-upon [*schlichte Erfassung*] to explication [*Explikation*], we find that it is a whole composed by parts that.

(i) have relations of interdependency, not only of foundation;
(ii) are submitted to global constraints that are not reducible to the individual properties of each one, but are instead properties of the whole as such;
(iii) are essential to the existence of the whole, so that the disappearance of several of them (or even one) entails the corruption of the whole itself; and
(iv) are highly sensitive to the environmental conditions, in a way that they do not subsist, nor the whole they form, if the environment changes dramatically.

When a whole is perceived as having these properties, then it is perceived as a concrete *natural* unity. The whole is not another piece put together with its parts. Nor is it reducible to the sum of the parts. It is a *moment*: the *form* of the wholeness or the pattern of its organization. There are aggregates which are not natural unities. But there are many natural wholes: a cell, a molecule of water, or complex, unpredictable systems like a water tourbillon or a hurricane.

This is what I have earlier referred to as the phenomenon of *self-organization.* It characterizes Nature itself, in the double sense that these units are "natural" and that Nature is the source of this kind of units. I call them "*forms*," without any compromise with the old-fashioned essentialism, because these forms are points of equilibrium between elements that are a function of the environmental conditions and vary if the environment varies. Given this dependency of forms on the "external" conditions, one could say in a strong sense that the only *absolute concretum* is Nature itself.

Thus, I agree with Seebohm when he emphasizes this kind of wholes. However, I disagree when he affirms that these self-organized wholes are only organic-wholes, i.e., living organisms. My hunch is that organization begins when Nature begins. Now the question is: how can we account for the phenomenon of organization as a *natural fact*? We must, obviously, give-up the bold and straightforward teleological concepts. We can no longer think as if the whole precedes and directs in advance the arrangement of the parts. Instead, in order to recognize organization as a strictly natural phenomenon and consider it as a self-organizing process, the following aspects are worth being considered:

Knotting, i.e., a tendency to join on the basis of some common or matching feature, such as, for instance, frequency, energy level, opposite charge, or pairs of opposite spins, among others: the entities resulting from knots can be "homeomeric" (the parts added produce only one entity of the same nature, like the superposition of waves in phase), or "heteromeric" (the parts joined produce a two-fold or *n*-fold entity, like opposite electron spins in an orbital, and so on).

Reinforcement, i.e., jointed physical entities strengthen the knots by mutually catalysing the elements necessary to each other, so that they form a whole that conserves them and constitute a new entity that interacts with the surrounding medium in a new global way; the systems formed by reinforcement can be stabilized systems or unstable systems that need to get in the environment ever new conditions in order to endure.

Economy or least dispense, i.e., the feature that knotting and reinforcement are favored because they reduce the losses that each entity in isolation would have through the interactions with the surrounding world, as if nature obeyed a principle of least action or as if nature chose according to a principle of the "good path"; instead, nature simply conserves the elements that attained the conditions for preserving themselves by knotting and reinforcing, while the others disappear or enter other wholes through a process of physical natural selection, so to speak.

An organized system is not a deterministic event of Nature. It entails stochastic processes at its base and, for that very reason, is not predictable in a full deterministic way. It is also highly dependent on the medium, in such a way that there are not pre-definite *forms* of organization, but precisely those that each medium-conditions allow at each moment. While "form," i.e., organization, is the phenomenon that reveals Nature, there are not fixed, pre-definite forms. A science of self-organization must deal with these non-linear processes and, thus, it has to overthrow the old Cartesian linear ways of thinking. Yet, only organized structures can last and remain in Nature, so that we can see the springing of *phýsis* as the coming about of organized systems in never-ending growing degrees of complexity. We attain then a pretty different conception of matter. Descartes and modern thinking, in general, conceived matter as something that was laying passively under universal and unbreakable laws. Now, the question amounts to seeing matter as an auto-poetic process. Of course, its name would no longer be "matter." In my opinion, a phenomenological philosophy of Nature should be centred on natural wholes (they have a peculiar formal ontology), instead of on the phenomenology of the constitution of the *res extensa*, *temporalis*, and *materialis*, that is to say, of a simple bulk of matter in time and space.

Husserl himself moved in this direction when he talked about a switch between a morphological contemplation of nature and an exact one, according to a change in "attitude" [*Einstellung*]. As he wrote: "Fundamentally different concepts of reality: this means correlatively: fundamentally different species of perceptions, of apperceptions. Anyhow, we speak with reason that we can consider 'the' world, and equally, in single, 'the' things, 'the' realities, as morphological and as exact" (2012, 261). However, he claimed that the morphological contemplation of Nature was "pre-scientific" and more suitable to the practical interests of life, as if a fundamental criticism of the modern sciences in its Cartesian foundations was not a possibility opened by this strong insight that Nature *is* morphological. A formal ontology of morphological unities, as wholes sustained by complex relations of interdependency and not analysable in ultimate elements according to relations of foundation, is a piece of theoretical thought that still waits for a development. It could deliver the net of essential laws that sustain the perceptual acquaintance with natural beings, in which the phenomenon of *internal organization* plays a decisive role.

9.5 The Puzzles of Quantum Physics: A Final Reflection

If I am right, classical modern physics captures only an abstract aspect of Nature when it idealizes the concrete natural wholes and its aggregates by means of exact essences suitable for a mathematical treatment. Nature is morphological, while physics deals with exact and formal essences that are abstract moments of natural wholes or of aggregates of natural wholes—space, time, mass, motion, and so on, are *attributes* of natural physical systems and not the natural systems themselves (perhaps, in the future there will be a mathematics of morphological essences—it would be a mathematics of complexity).

Thus, regarding the mathematical science of Nature, one must distinguish, within its noematic content, between (i) the mathematical framework as a *mode of access* and (ii) the actual *determinations* of the physical object as such. In addition, one must distinguish between the noematic sense and the object referred to. Thus, the noematic sense is a complex construction of exact and formal essences, along with a theoretical conception of the reality to which it applies (a suitable determination of the essence of the region-Nature). The object referred to is the physical reality that (1) is conceivable and accessible precisely through this complex content, and (2) is directly or indirectly observable through the addition of measuring instruments. However, the noematic content is not the object, so that one cannot submit the former to reification and blend together the physical object and the mathematical framework that belongs to its "modes of access." In my opinion, Seebohm's insights go precisely in this direction: firstly, when he stresses that the departure from the lifeworld in the modern science of Nature was both triggered and conveyed by the development of mathematical formalisms; secondly, when he refers to the objects of contemporary physics with the expression "mathema-physical." Indeed, this last denomination captures in a single word the entire problem of the positing of a reality (the "physical") whose access modes are developed in the realm of exact and formal essences. There are two possibilities: *either* the object is pre-given in the lifeworld, so that the physical-mathematical framework and the correlated instruments of measurement determine only some aspects of it (like "mass," "momentum," "magnetic field," and so on, for the macroscopic realities thematically considered as "bodies" in the perceptual world); *or* the mathematical framework exhausts the "modes of access" (there is no perceptive acquaintance) and determines the "modes of givenness" of the object (observable quantities through apparatuses), so that the object is no longer accessed and given independently from them. This last case is the particular situation of contemporary physics, and more specifically of quantum mechanics, that the concept of "mathema-physical" objects tries to capture.

In quantum physics, what Husserl calls "the object in the mode of its determinations" [*Gegenstand im Wie seiner Bestimmtheiten*] would be only the "observables" that allow a glimpse into the underlying physical reality that is posited in advance. They are the *definite* eigenstates of a quantum entity (a definite position, a definite spin, and so on). The eigenvectors, the Hermitian operators, its underlying algebra and the Hilbert spaces belong to the logical-mathematical structure of the noematic content, not to the object which is intended by it. This mathematical framework "subsists" as an ideality with no real existence. The intentionality that goes to it is not the intentionality that intends something with the sense of a *physical* reality. Certainly, the physical object only appears as *measured* in the framework of the mathematical formalism—as against the objects that are given in the lifeworld, it has no other mode of access. Nevertheless, the mathematical formalism is not a horizon from which the object detaches itself, in the way the co-givenness of the world furnishes an external horizon for the givenness of each individual object. Thus, there is here only a partial analogy with classical physics. On the one hand, for instance in mechanics, the physical is some sudden, determinate variation of the velocity of a body, while the second derivative of the function in respect to time, through which the acceleration is measured, belongs not to the physical proper, but rather to its mode of access. On the other hand, what precludes the blending of the mathematical framework and the physical object is the fact that the concepts of classical physics are idealizations that refer back to real entities in the perceptual world. They are exact essences, which, by themselves, allow a pictorial representation and eventually a reference to bodies in the perceptive surroundings. Thus, here there is no possibility of an overlapping between the mathematical, on the one hand, who defines a mode of access, and, on the other hand, the physical reality itself that is intended to (the *Gegenstand schlechthin*) plus the physical object in the mode of its determination (the *Gegenstand im Wie*; here, a change in the velocity rate). Apparently, this is not the case in quantum physics. In accordance, in the Copenhagen interpretation we find some mathematical objects (such as the superposition of states in a Hilbert's probability space) construed as physical realities. The physical is here engulfed by the mathematical formalism, and the formalism allows only a pictorial representation *per analogiam*. As a consequence, there is no reality in the lifeworld that can be exhibited as the object to which the formalism ultimately refers. Because of that, the mathematical formalism is no longer taken as a mode of access to a pre-given object; it is rather posited as the object itself. Additionally, when the claim of completeness is accepted, what the mathematical formalism describes is not only the physical reality itself, but *all* that there is in the quantum realm. The strange situation is that the supposition of an underlying and independent reality is no longer allowed in this construal. This adds to the strange view that definite states of the dynamical variables of a system are only achieved by means of an observation, which supposedly triggers the "collapse" of the wave-function. Thus, not only the position of a reality recedes into the mathematical formalism, but also the physical only gets a determinate value when there is an observation. In a sense, there is no quantum reality in the pregnant sense when "no one is looking," only a conflation of mutually incompatible possible states. Seebohm refers to this physical-observational blend, opposing it to the de Broglie-Bohm alternative version, where such strange conflation between the physical and the mathematical is avoided. In my opinion, this kind of interpretation in "Copenhagen spirit" belongs to the oddities a phenomenological ontology of Nature would help to criticize.

This lack of an organized ontology in the Copenhagen interpretation is flagrant in Bohr's principle of complementarity. Indeed, it tells us the following disappointing story:

1. Quantum entities *show* us nothing by themselves.
2. If proved by an apparatus, quantum entities can *manifest* wave-like behavior or corpuscle-like behavior.
3. The behaviors of quantum entities are always behaviors *relative to* and *in the context* of a definite observational situation.

4. These corpuscle-like and wave-like behaviors are not manifested simultaneously and by the interaction with the same apparatus.
5. For that reason, the concepts of corpuscle and wave do not contradict each other, given that they are not applied at the same time and in the same experimental context to the same quantum entity.
6. Instead, they exclude each other, but are both jointly necessary to the description of the diversity of quantum manifestations.
7. So, firstly, we have no grips on the quantum entities, but only the possibility of applying to them concepts we have borrowed previously from the macroscopic world of classical physics.
8. Secondly—as it seems Bohr himself has put it—for us there is no such thing as a quantum world, but only a quantum physical description.
9. And, thirdly, quantum entities are well-founded, although limited, conjectures from the macroscopic (classical) world into the microscopic (non-classical) world, which return to some processes and phenomena in the macroscopic (classical) world (e.g., the classical concept of a corpuscle gives rise to the quantum concept of an electron movement, which serves to interpret a visible trace of condensed water vapor in an ionization chamber).

We see that complementarity puts severe limitations concerning not only the possibility of conjoining causal description and space-time coordination, but also regarding the very possibility of taking quantum phenomena as a realm of real entities, i.e., of taking them as appearances of an *actual world* of things and events. As a matter of fact, the quantum phenomenon, as described in Bohr's and Heisenberg's interpretations, lacks the conditions to be considered an *entity* in the full sense of the word: it has no independence from the measuring device, no integrated description of its own multiple manifestations, no concepts borrowed in its very nature, and no such fundamental things as continuity across space-time, determined actual states and separation. And so we face, with the Copenhagen interpretation, the following devastating situation for our hopes to understand Nature: fundamental concepts are not *realistically* interpretable. That is to say, the very concepts of physics we are using do not allow us to elaborate a sound and coherent representation of *phýsis*.

It is one of the strange things of the last century that the model put forth by Louis de Broglie and further developed by Bohm has remained at the margins of quantum physics, while the so-called "Copenhagen spirit" remained almost dominant. We are told by professional physicists that they are mathematically equivalent, and then that there are no criteria available for a decision. Nevertheless, this argument does not justify the manifest penchant the majority of professional physicists have for the Copenhagen interpretation. I think that this is the right place for the intervention of a phenomenological philosophy of Nature, and not simply for a phenomenological epistemology of physics accepting the way physicists usually interpret the ontological import of their professional work. Obviously, this intervention will not settle the debate among physicists. However, the huge questions that are at stake cannot be indifferent for the philosopher, because these questions are not strictly physical in a technical sense. They are ontological in a strong sense, so that the philosopher cannot withdraw from them and remain apart.

The core of the de Broglian interpretation lies in considering every quantum entity as a complex system constituted by a pilot-wave and an energetic peak that can be treated as particle-like. The behavior of a quantum entity (like a photon, an electron, etc.) is determined by the interactions between the "particle" and its wave, not considering the dependencies with other entities. In addition, the so-called "empty-wave" is simply a mathematical construction that *models* the real pilot-wave in which the particle is entrenched. So, there is no conflation here between the physical and the mathematical. Besides, instead of a "reality" that is only definite when observed, here we find real trajectories, not mysterious "collapses," and a sound, logical explanation for the results of the double-slit experiment that is free from paradoxes. Not surprisingly, the quantum-realm restores significant links with the physics of macroscopic systems, namely with hydro-dynamic systems, when it is conceived as a *real complex system* composed of a wave interacting with its "particle." As it was recently shown by several experiments with droplets interacting with a vibrating fluid, the behavior of a quantum entity has strong analogies with the behavior of these macroscopic systems. The behavior of droplets can mimic phenomena supposed to be specific of quantum physics, like particle diffraction, quantum tunneling, quantized orbits, the Zeeman Effect, interference patterns (Young's experiment), and the quantum corral (Bush 2015). All in all, while maintaining a great difference between the two realms (classical and quantum systems), like non-locality, we are definitely allowed to talk about a physical realm in a realistic sense, i.e., of a quantum-world that has definite states *in itself* and then states *relative* to an observer, inasmuch as observation at so small a scale always disturbs the system which is observed. Nevertheless, there is a system *before* observation, albeit we only get the results of its interaction with our apparatuses.

For a philosophy of Nature, this is very important. Nature has a morphological character, in the sense I stressed above of a "concrete whole." The mathematical framework is able to deal with Nature when and only when its objects are idealized and the morphological wholes (which are complex) are treated by means of exact essences (which are suitable for a linear treatment). Nevertheless, these idealized objects (mass, space, acceleration, etc.) are not the physical systems, but only abstract moments of them. When there is no direct access to the morphological wholes, when they are not pre-given in the lifeworld, the tendency arises for blending the physical and the mathematical framework, as if there was no objective world but only a "quantum-mathematical" description, as Bohr seems to have put it. In my view, the rupture between the classical and post-classical formalism has to do with the fact that quantum physics is a physics of the "invisible," in the sense that its theoretical objects cannot eventually be referred to morphological entities in the perceptual world. Nevertheless, this situation does not prevent us to conceive of a morphological system and posit it as the *independent reality* we are searching for. This is the case with the so-called de Broglie/Bohm approach. Fortunately, there is an increasing amount of experimental evidence that endorses it. In my view, this is the path for a new productive encounter between the science of Nature and a philosophy of Nature that is very far from the reductionist approach that prevailed in positivism and analytic philosophy. Of course, I am referring to phenomenology.

I finish with a quote from Husserl's *Umarbeitung* for the *Crisis*-book:

> In the abstract stratum of "corporality", the lifeworld [...] becomes a "world" to which there is an exact foresight, it becomes a calculable world, certainly in an infinite progress.
>
> ... However, in the tradition, it has never been realized that it was a mere abstraction, and that a Nature in the sense of the Science of Nature is at all unconceivable as a world of "bare bodies." With Physics, we do not get a concrete knowledge of the world. There are no bodies that would be simply physical bodies [in the sense of the mathematical Physics]. The world of concrete things has an all-structure, and certainly the things as its concrete. [...] Thus, science has the entire task in the world as world of concrete. [...] According to our view, certainly [mathematical] Physics has no region in a concrete sense. (1992, 147)

References

Bohr, Niels. 1928. The quantum postulate and the recent development of atomic theory. *Nature* 121: 580–590.

Born, Max. 1926. Zur Quantenmechanik der Stoßvorgänge. *Zeitschrift für Physik* 37: 863–867.

Bush, John W.M. 2015. Pilot-wave hydrodynamics. *Annual Review of Fluid Mechanics* 47: 269–292. (review paper).

De Broglie, Louis. 1924. *Recherches sur la Théorie des Quanta*. PhD dissertation. *Annales De Physique* 10eme Série, t. III (jan-fév. 1925).

Hardy, Lee, ed. 2013. *Nature's suit: Husserl's phenomenological philosophy of the physical sciences*. Athens: Ohio University Press.

Husserl, Edmund. 1977. *Ideen zu einer reinen Phänomenologie und phänomenologischen Philosophie, Erstes Buch: Allgemeine Einführung in die reine Phänomenologie 1. Halbband: Text der 1.-3. Auflage*, Husserliana III-1. ed. Karl Schuhmann. The Hague: Martinus Nijhoff. Reprint.

———. 1992. *Die Krisis der europäischen Wissenschaften und die transzendentale Phänomenologie. Ergänzungsband. Texte aus dem Nachlass 1934–1937*, Husserliana XXIX. Dordrecht: Kluwer Academic Publishers.

———. 2002. *Natur und Geist. Vorlesungen Sommersemester 1919*, Husserliansa-Materialen V. Dordrecht: Kluwer Academic Publishers.

———. 2012. *Zur Lehre vom Wesen und zur Methode der Eidetischen Variation*, Husserliana XLI. Dordrecht: Springer.

Seebohm, Thomas. 2015. *History as a science and the system of the sciences: Phenomenological investigations*. New York: Springer.

Pedro M. S. Alves is Professor of Philosophy at the University of Lisbon and Director of the Line-Research of Phenomenological Thinking at the Center of Philosophy at the same university. His main interests are theory of knowledge, phenomenology, and the conceptual foundations of physics. He published several books and papers on phenomenology. He is also editor of *Phainomenon*, an editorial project comprising a phenomenological journal, a collection of monographic studies, and a series of translations into Portuguese of the classics of phenomenology, especially Husserl's works.

Chapter 10
The Inadequacy of Husserlian Formal Mereology for the Regional Ontology of Chemical Wholes

Marina Paola Banchetti-Robino

Abstract In his book, *History as a Science and the System of the Sciences*, Thomas Seebohm articulates the view that history can serve to mediate between the sciences of explanation and the sciences of interpretation, that is, between the natural sciences and the human sciences. Among other things, Seebohm analyzes history from a phenomenological perspective to reveal the material foundations of the historical human sciences in the lifeworld. As a preliminary to his analyses, Seebohm examines the formal and material presuppositions of phenomenological epistemology, as well as the emergence of the human sciences and the traditional distinctions and divisions that are made between the natural and the human sciences.

As part of this examination, Seebohm devotes a section to discussing Husserl's formal mereology because he understands that a reflective analysis of the foundations of the historical sciences requires a reflective analysis of the objects of the historical sciences, that is, of concrete organic wholes (i.e., social groups) and of their parts. Seebohm concludes that Husserl's mereological ontology needs to be altered with regard to the historical sciences because the relations between organic wholes and their parts are not summative relations. Seebohm's conclusion is relevant for the issue of the reducibility of organic wholes such as social groups to their parts and for the issue of the reducibility of the historical sciences to the lower-order sciences, that is, to the sciences concerned with lower-order ontologies.

In this paper, I propose to extend Seebohm's conclusion to the ontology of chemical wholes as object of quantum chemistry and to argue that Husserl's formal mereology is descriptively inadequate for this regional ontology as well. This may seem surprising at first, since the objects studied by quantum chemists are not organic wholes. However, my discussion of atoms and molecules as they are understood in quantum chemistry will show that Husserl's classical summative and extensional mereology does not accurately capture the relations between chemical wholes and

M. P. Banchetti-Robino (✉)
Florida Atlantic University, Boca Raton, FL, USA
e-mail: banchett@fau.edu

© Springer Nature Switzerland AG 2020
T. Nenon (ed.), *Thomas Seebohm on the Foundations of the Sciences*,
Contributions to Phenomenology 105,
https://doi.org/10.1007/978-3-030-23661-8_10

their parts. This conclusion is relevant for the question of the reducibility of chemical wholes to their parts and of the reducibility of chemistry to physics, issues that have been of central importance within the philosophy of chemistry for the past several decades.

10.1 Introduction

In his book, *History as a Science and the System of the Sciences*, Thomas Seebohm articulates the view that history can serve to mediate between the sciences of explanation and the sciences of interpretation, that is, between the natural sciences and the human sciences. Among other things, Seebohm analyzes history from a phenomenological perspective to reveal the material foundations of the historical human sciences in the lifeworld.[1] As a preliminary to his analyses, Seebohm examines the formal and material presuppositions of phenomenological epistemology, as well as the emergence of the human sciences and the traditional distinctions and divisions that are made between the natural and the human sciences.

In his examination of the formal methodological presuppositions of phenomenological epistemology, Seebohm begins by outlining the history of the development of the human sciences and by considering the dominance of psychologism in the late nineteenth and early twentieth centuries. One of the paradoxes of psychologism is that the epistemology of the formal sciences, that is, of the sciences concerned with ideal objects, would have to become a branch of empirical science. As Seebohm points out, naturalistic psychologism chooses to live with this paradox. Edmund Husserl, on the other hand, rejects this psychologistic approach, since it would entail that mathematics should become a branch of empirical psychology. Instead, Husserl chooses to apply phenomenological descriptions immediately to epistemology. In fact, the second volume of the *Logical Investigations* directly concerns the application of phenomenological descriptions to epistemology. As Seebohm notes, Husserl seeks to demonstrate that "the assumption that phenomenology can be applied in epistemological investigation about the formal sciences and, in general, to ideal objects implies that phenomenology cannot be understood as descriptive psychology" (2015, 14). The relevance of Husserl's task for Seebohm's thesis is that Husserl shows that "phenomenology is in this phase of its development the method not only of epistemology, but also of all other disciplines of philosophy" (14). Thus, just as Husserl's phenomenological epistemology reveals the foundations of the formal sciences within eidetic structures, Seebohm seeks to show that a reflective phenomenological attitude with respect to the historical sciences will reveal their material foundations in the lifeworld.

---

[1]This paper is dedicated to the memory of Professor Lester Embree, who was my respected colleague of many years and whose advice and support were always generous and invaluable to the progress of my work.

In his discussion of Husserl's application of phenomenology to the epistemology of the formal sciences, Seebohm devotes a section of his book to Husserl's theory of parts and wholes, that is, to mereology, as it is developed in the second volume of the *Logical Investigations* and later in *Experience and Judgment*.

> Understood as a general theory of parts and wholes, mereology has a long history that can be traced back to the early days of philosophy. As a formal theory of the part-whole relation—or rather, as a theory of the relations of part to whole and of part to part within a whole—it is relatively recent and came to us mainly through the writings of Edmund Husserl and Stanisław Leśniewski. [Husserl's work was] part of a larger project aimed at the development of a general framework for formal ontology … [and it] finds its fullest formulation in the [second volume] of his *Logical Investigations*. (Gruszczyński and Varzi 2015, 409–410)

For Husserl, ontology is to be understood in the old sense of a theory of being and "its task is to lay bare the formal structure of what there is *no matter what it is* … For instance, it would pertain to the task of formal ontology to assert that every entity, no matter what it is, is governed by certain laws concerning identity, such as reflexivity, symmetry, or transitivity" (410). Identity, reflexivity, and transitivity are formal relations since they may be said to apply to anything that might exist, and Husserl believed that the part-whole relation is also formal in this sense. "Parthood seems to apply to entities as different as material bodies (the handle is part of the mug), events (the first act is part of the play), geometrical entities (the point is part of the line), etc. Even abstract entities such as sets, appear to be amenable to mereological treatment" (411). Thus, Husserl believed that "a purely formal theory of wholes and parts was possible" (Simons 1982, 114) and would serve as the foundation of the mereology of regional ontologies.

In fact, Seebohm devotes a section to discussing Husserl's formal mereology precisely because he understands that a reflective analysis of the foundations of the historical sciences requires a reflective analysis of the objects of the historical sciences, that is, of concrete organic wholes (i.e., social groups) and of their parts. Thus, Seebohm is first concerned with discussing Husserl's mereological ontology of the formal sciences and then with examining whether this extensional mereology suffices to properly describe the relations between the organic wholes and their parts that form the objects of the historical sciences. Seebohm concludes that Husserl's mereological ontology needs to be altered with regard to the historical sciences because the relations between organic wholes and their parts are not the kinds of summative relations that exist between, for example, mugs and their parts or sets and their parts. Seebohm's conclusion is relevant for the issue of the reducibility of organic wholes such as social groups to their parts and for the issue of the reducibility of the historical sciences to the lower-order sciences, that is, to the sciences concerned with lower-order ontologies.

In this paper, I propose to extend Seebohm's conclusion to the ontology of chemical wholes as objects of quantum chemistry and to argue that Husserl's formal mereology is descriptively inadequate for this regional ontology as well. This may seem surprising at first, since the objects studied by quantum chemists are not organic wholes. However, my discussion of atoms and molecules as they are understood in quantum chemistry will show that Husserl's classical summative and extensional mereology does not accurately capture the relations between chemical wholes and their parts.

I will begin by examining the principles of Husserlian mereology for formal ontology, then focus on Seebohm's critique of the applicability of Husserlian mereology to the higher-order regional ontologies of social groups, and I will conclude by arguing that this same critique can be extended to the lower-order regional ontology of chemical wholes and parts. To support my arguments, I will discuss concrete examples from the field of quantum chemistry, albeit in a non-technical manner, to show that the summative extensional mereology developed in the *Logical Investigation* and *Experience and Judgment* cannot successfully capture the distinctive relations that exist between molecules and their atomic parts. Again, this conclusion is relevant for the question of the reducibility of chemical wholes to their parts and of the reduction of chemistry to physics, issues that have been of central importance within the philosophy of chemistry for the past several decades.

10.2 The Principles of Husserlian Mereology for Formal Ontology

Husserl devotes a generous part of the second volume of the *Logical Investigations* and, later, of *Experience and Judgment* to the theory of wholes and parts because he considers mereology as fundamental to the phenomenology of the formal sciences. This is because he understands pure logical grammar as "the theory of apophantic forms of complex independent unified wholes of meaning, *Bedeutungskomplexionen*" (Seebohm 2015, 16). Seebohm points out that, certainly, the formal ontological theory of wholes and parts that is developed in the *Logical Investigations* suffices for the analysis of apophantic forms and of "the hyletic correlates of passive synthesis in primordial sensual experience," (19) what he refers to as first-order wholes. In the *Logical Investigations*, Husserl concludes that the formal ontology of parts and wholes is analogous to that of units and manifolds so that, just as manifolds are additive collections of units, wholes are analyzed as additive collections of parts (see Simons 1982).

For our discussion, it is important to note that one significant aspect of the Husserlian theory of parts and wholes is his extensive discussion of the notions of dependent parts (moments) and independent parts (pieces).

> *Each part that is independent relatively to a whole W we call a Piece (Portion), each part that is non-independent relatively to W we call a Moment (an abstract part) of this same whole W.* It makes no difference here whether the whole itself, considered absolutely, or in relation to a higher whole, is independent or not. *Abstract parts can in their turn have pieces, and pieces in their turn abstract parts.* (Husserl 1970, §17, 29)

This aspect of Husserl's theory is crucial for our discussion because, to answer the question of the reduction of a higher-order ontology to a lower-order ontology, one must examine whether the parts of the wholes that constitute the objects at the higher level are pieces of the whole (in which case reduction is materially possible) or moments of the whole (in which case reduction is not materially possible). This is the case because, although pieces of wholes can exist independently of the whole of which they are independent parts, moments of wholes cannot exist independently of the whole of which they are dependent parts. One possible example of this distinction is given by Peter Simons: "The board which makes up the top of the table is a piece of the table [while] the surface of the table, or its particular individual color-aspect, are moments of it" (1982, 115). Husserl believes that this distinction can be extended from the psychological sphere to objects in general, which is why it becomes a general component of his formal ontology. Husserl also specifies that he is using the term "part" in a wide sense that includes not only detachable pieces of wholes but also "anything that is a constituent of it, apart from relational characteristics [including] so-called accidents and also boundaries" (120).

Husserl distinguishes three different sorts of wholes and this distinction is also important in our consideration of reductionism. The first sort is a whole considered in the narrow sense or what Barry Smith calls a "narrow" whole (Smith and Mulligan 1982, 121). This is one in which entities have come together by virtue of the kinds of entities that they are, without anything else joining them together apart from a "unifying moment," which is itself not another entity.[2] Are the entities that come together to form "narrow" wholes to be considered pieces or moments of the whole? From a formal perspective, the existence of such entities is independent of the whole because their existence as the kinds of entities they are is not dependent upon their existence within the whole. Instead, it is their intrinsic nature as certain sorts of individuals that necessarily binds them together to form the whole, which is extrinsically unified. So, the answer to the question above would have to be that the parts of "narrow" wholes are to be considered as pieces of the whole, rather than as moments (Husserl 1970, §22–24, 37–40).

The second sort is a whole considered in the "wide" sense or what Smith calls a "wide" whole, which can be regarded as a single thing, regardless of how scattered its parts are or how tightly or loosely connected they are. This allows the inclusion both of unities and pluralities, since "any plurality may be taken together as something unitary, thereby founding a new higher unity, whose unity is, however, extrinsic to it, in the collective act" (Simons 1982, 122). In this case, it is obvious that the parts of a wider whole are pieces, that is, independent parts of the unity or plurality of which they are parts. In fact, what is regarded as a "wide" whole seems to be a factor of some decision to trace the boundaries of the whole at one place rather than another, whether the decision is arbitrarily made or not. As Simons points out, one can "allow as individuals anything which can possess a (singular) proper name. This will include arbitrary *collectiva*. This liberality is reflected in extensional mereologies by allowing that arbitrary sums of individuals are themselves individuals.

[2] Husserl is here trying to avoid the regress, discussed by F. H. Bradley in *Appearance and Reality*, that is generated when one requires a third entity to join the first two so that: "if A and B are bound together by U, then A and U must be bound together by U[1], and so on *ad infinitum*" (Simons 1982, 121).

The reason for this is … that it is not clear in advance where to draw the line between things which are wholes in this widest and weakest sense, and those which have some more intrinsic unity"(122).

The third and more interesting sort of whole is a "pregnant" whole, which Husserl defines as "a range of contents which are all covered *by a single foundation* without the help of further contents. The contents of such a range we call its parts. Talk of the *singleness of the foundation implies that every content is foundationally connected whether directly or indirectly, with every content* … without external assistance" (Husserl 1970, §21, 34). Unlike mere sums or aggregates ("narrow" and "wide" wholes) whose unity is extrinsic, the unity of pregnant wholes is intrinsic, but the parts of such wholes are still to be considered as pieces. It is the relations between those parts that are, for Husserl, to be considered as moments. Such relations can be either ideal or real relations. Ideal relations are such that they do not alter their terms in any way nor do they bring their terms into any real connection to each other. For example, if A is the same height as B, both A and B remain unaffected by their being in this relation to each other. Ideal relations are often equivalence relations and are, thus, reflexive, which is why ideal relations cannot "engender genuine foundation relations" (Simons 1982, 154).[3] Real relations, on the other hand, are such that they alter or affect their terms in some way. For example, if magnet A attracts metal B, both A and B are affected by this magnetic attraction. Thus, the most obvious example of a real relation is one that involves a causal link (154). Real relations are clearly not reflexive and can serve to engender genuine foundation relations. When they do so, such relations can be described as moments of the whole that unite the parts (154). Some foundation relations are symmetric (*two-sided* or *mutual* foundations), such as the foundation relation between color and extension which are mutually founding. Other foundation relations are asymmetric (*one-sided*), such as the foundation relation between a lake and the dry land upon which the lake is one-sidedly founded.[4]

These are the three types of wholes discussed by Husserl within the context of formal ontology. However, if ontology is to be understood in the old sense of a theory of being whose "task is to lay bare the formal structure of what there is *no matter what it is*" (Gruszczyński and Varzi 2015, 410) and if mereology reveals the fundamental principles of such formal structure, the question is what constitutes natural wholes and whether the mereology of the *Logical Investigations* and *Experience and Judgment* suffices to adequately describe the whole-parts relations of the categorial objects within the regional ontological spheres of the material world. When we are considering natural wholes or systems, it is the causal integrity of such systems that binds them together (Simons 1982, 150), and we may ask

[3] Simons regards the whole notion of a relation that holds between a thing and itself as suspect, especially in cases such as identity which can only hold between a thing and itself. Thus, Simons considers reflexive relative terms to be "ontologically sterile" since "nothing intrinsically relational is represented" by such terms.

[4] The terms "narrow" whole and "wide" whole are coined by Barry Smith, while "pregnant" whole is Husserl's own term.

whether the parts of such wholes are to be considered as independent parts (pieces) or as dependent parts (moments). Husserl regards objects in the world as being naturally organized in many different ways, so that the distinction between lower-order and higher-order objects is not absolute. From one point of view, an object may be "a natural unit, from another it may coincide with an aggregate of differently organized units, or again be a moment of a greater whole" (150). However, the relativity described here "is not the mere imposition of a conceptual scheme on an otherwise unstructured world, but cuts along natural seams in reality." It would follow from this that, at least for Husserl, the parts of even higher-order wholes are to be considered pieces rather than moments. Yet, both Seebohm's discussion of organic wholes and my discussion of chemical wholes will call this into question, since Seebohm challenges the idea that the parts of organic wholes are independent parts, while I will challenge the idea that the parts of chemical wholes are independent parts.

As Rafał Gruszczyński and Achille Varzi have recently pointed out, "the general applicability of the part-whole relation is controversial … [David] Lewis himself famously argued that entities such as universals *cannot* be structured mereologically, short of unintelligibility" (2015, 409–10). To illustrate this point Lewis, interestingly, uses an example taken from chemistry: "Each methane molecule has not one hydrogen atom but four. So, if the structural universal methane is to be an isomorph of the molecules that are its instances, it must have the universal hydrogen as a part not just once, but four times over… . But what can it mean for something to have a part four times over?" (Lewis 1986, 34). This difficulty concerns the bounds of mereology as formal ontology, but there are also other difficulties that concern its content. "Consider, for instance, the question of whether there are mereological *atoms* (i.e., entities with no proper parts), or of whether everything is ultimately *composed* of atoms. Clearly any answer to such questions would amount to a substantive metaphysical thesis that goes beyond a 'pure theory of objects as such.'" (Gruszczyński and Varzi 2015, 412). In the words of Gruszczyński and Varzi, "taken together, then, these two sorts of difficulty represent a serious challenge to the idea that mereology can form a genuine piece of formal ontology" (413).

This paper will not be concerned with this second difficulty but will focus only on the first, that is, does Husserl's conception of the parts-whole relation as formal ontology extend to all regional ontologies as he believed? The paper will argue that the answer to this question is negative to the extent that the formal principles of Husserlian mereology do not, in fact, extend either to the higher-order ontologies of organic wholes discussed by Seebohm or to the lower-order regional ontology of chemical wholes as these are conceived in quantum chemistry. I will first examine Seebohm's arguments regarding organic wholes, then I will examine chemical wholes as they are conceptualized but also manipulated and transformed by chemists. Here, I will show that, for the purposes of quantum chemical calculations, chemists must conceptualize subatomic "entities" in ways that violate the principles of standard mereology such as transitivity, unrestricted composition, and uniqueness of composition. For example, Kit Fine discusses a fundamental principle of standard mereology, with which Husserl would agree, according to which "the same parts cannot, through different methods of composition, yield different wholes"

(1994, 138). That is, according to this principle, the same composition of parts structured differently will yield the same whole, with the same properties. As well, Lewis explains another fundamental principle, that is, although any two or more individuals constitute a mereological sum or fusion, inclusion in such a sum or fusion does not alter the individuals that are included (1986, 25–26). The mereological sum or fusion has no causal effectiveness as such or as separate from the causal effectiveness of its parts.

As we shall see in this essay, however, it is precisely these aspects of classical extensional mereology that render it inadequate for describing the whole-parts relations both of organic wholes but also of chemical structures as understood in quantum chemistry. I now turn to Seebohm's discussion of the whole-parts relations for the higher-order ontologies of organic wholes as it appears in his book *History as a Science and the System of the Sciences*.

10.3 The Application of Husserlian Mereology to the Higher-Order Regional Ontologies of Organic Wholes

As Seebohm points out, the mereological theory of the *Logical Investigations* and *Experience and Judgment* is a "general ontological theory of wholes and their parts" (2015, 20) that presupposes an abstractive reduction that focuses on first-order wholes with parts that are "held together by unifying foundations" (20), in other words, pregnant wholes. However, there are more complex types of wholes and parts that are referred to in natural language, what Seebohm calls "wholes of a higher order" (19). Among higher-order wholes are included the solar system, organisms of varying degrees of complexity, and social groups and communities. Such wholes are of a higher order precisely because their constituent parts are themselves wholes rather than simple parts. However, Husserl does not offer a formal account of such structures, precisely because he is only concerned with developing a mereology for the formal ontology of pure logical grammar, not with the mereologies of categorial objects of a higher order. Seebohm raises the question of "whether some types of such categorial objects of a higher order can have formal ontological structures that count as the formal structures of the wholes of a higher order" (20). Certainly, as Seebohm points out, Husserl's discussion in the *Logical Investigations* begins in the psychological sphere but is intended to be valid and applicable to all objects, psychological or not, because it appeals only to properly generalizable features.

For Seebohm, then, the question is really "whether such a theory can admit additional categorial structures for wholes with parts [that are not parts of a unifying system of *foundations* but] that are already themselves wholes connected by unifying systems of relations" (20). As he emphasizes, "going beyond *Experience and Judgment* it can be said that the differences between collections as categorial forms of a higher order are determined by structured systems of relations" (20). In a series of specific comments on this point, Seebohm proposes that the criterion for demarcating "between the unifying systems of parts in wholes of the first order and wholes of higher order is the difference between foundations and relations. Foundations share some of the formal properties of relations, but foundations are 'relations' between dependent moments that cannot be given only in the context with other moments and between pieces as relative independent parts of wholes" (21). To clarify this passage, foundations concern parts that can be given independently of the context with other moments. On the other hand, relations as they pertain to the formal ontological theory of categorial structures of higher-order wholes are "relations between independent wholes" (22). In other words, we are not talking here about relations between n-adic predicates that refer to individual objects but about relations between parts that are themselves independent wholes.

There is another distinction that needs to be made, however, when one considers the mereology of higher-order wholes. This is the distinction between inorganic and organic wholes of a higher order, and it is a distinction that is extremely relevant when one reflects upon those wholes that are the objects of the life sciences and of the human and social sciences. Seebohm clarifies that phenomenological epistemology is not concerned with answering the empirical question of whether and of how organic wholes can emerge from inorganic matter. Rather, phenomenological epistemology

> is restricted to the analysis of the cognitive attitude of the life sciences and their intentional correlate, the ontological region of organic entities. It includes in addition the analysis of the relations between the categories of the ontological region of organic life and the categories of the ontological region of inorganic matter. Such analyses are, however, able to decide the question of whether the reduction of the life sciences to the hard sciences is an ideal formal ontological possibility and then a material ontological possibility. This will be the case if it can be shown that the ontological region of inorganic matter is the static and genetic foundation of the ontological region of organisms. (242)

Regarding the phenomenological epistemology of the life sciences, Seebohm establishes that it is not enough to simply analyze the formal categorial structures of the formal ontology of the wholes and parts. One must also analyze the formal ontological theory of unit and manifold and establish what the relation is between the theory of wholes and parts and the theory of units and manifold (242–243). "*Either* they are two independent formal ontological theories on the same level of universality *or* one of them belongs to a higher logical level of universality … It is obvious that the reduction of organic life to inorganic matter is a priori a formal ontological impossibility if the answer is 'yes' to the first horn of the dilemma and 'no' to the second. The reduction is, however, an ideal possibility if the answer is 'no' to the first and 'yes' to the second horn of the dilemma" (243). The question, therefore, really turns on whether organic wholes can be treated like collections and, therefore, analyzed as manifolds, which are additive collections of units. If so, then one can answer yes to the first horn of the dilemma and no to the second.

According to Seebohm, the latter is in fact the case so that reduction is an ideal possibility. He states that

> Measuring as counting of units and numbers as units in measuring can be applied to all dependent parts that belong to the genus extension. But this means that all entities under the formal ontological categories of the whole and the parts can be considered as units and collections of units if the formal ontological differences pertaining to the categories of the theory of the whole and parts are excluded with the aid of a generalizing abstraction determining a realm in which all of them, including the wholes themselves, are mathematical units in collection that can be themselves considered as units, etc. . . . The formal ontological theory of unit and manifolds is, hence, one-sidedly founding for the formal ontological category of the whole and the parts. Manifolds of concrete wholes as well as independent parts can be counted. A reduction of the material ontological structures of organic entities is, hence, an ideal possibility. (243–244)

Lest the reductionist declare victory at this point, Seebohm stresses that a reduction of the material ontological structures of organic entities is "only an *ideal* possibility [and not a material one] because a genetic foundation of B in A requires, beyond A, the additional structures of properties C for B" (244). The additional structures of properties C for B are discovered by carrying out an analysis of the structures of the material regional ontology to which B belongs. Thus, in the case of organic wholes, "the second dimension of the description of phenomena that are necessary for the explication of the material categories of organic life and organisms has to determine the material characteristics of the parts of organic wholes. The independent parts of organic wholes cannot be simultaneously parts of other organic wholes, and they cannot exist independently outside the system of their functions in the organic whole. They will decay if they are separated from the wholes without providing an artificial environment that can substitute for the whole or stop the process of decay" (245).

Seebohm also says something very important about the role that an analysis of environment must play in the phenomenological epistemological analysis of organic wholes.[5] He states that "the description of the environment of a species and the ecological relation of the species to the environment is not of immediate significance for the simple recognition of the species of an individual organic being or for the development of taxonomies of organic species. It is, however, of significance for experimental research and for the discovery of causal relations between organisms as concrete whole (or their parts) and certain properties or aspects of their environments" (245). Thus, although a description of the environment is not part of the first dimension of the description of organic wholes, it is crucial for the second dimension of the phenomenological description since it pertains to the structures of properties C for B that are discovered through an analysis of the material regional

[5] This attention to the relation between organism and environment was emphasized in the late nineteenth century by Louis-Adolphe Bertillon, who coined the term mesology to refer to what he called "*la science des milieux*," a concept that he later expanded to the field of sociology. The focus on mesology has recently been resurrected by Augustin Berque to, once again, call attention to the importance of milieu in the study of organisms and as part of a proposal for overcoming modernist reductionism. See Berque 2010.

ontology of B. Ultimately, however, the material ontological categories of organic wholes are … in the last instance one-sidedly founded in the categorical system of the ontological region of the hard sciences" (248), that is, physics and chemistry. But, this "does not mean that organic life can be *reduced* to the system of the categories of the hard sciences. What is again in question is the factor C that must be added to the founding material ontological region A for the emergence of entities belonging to the material ontological region B" (249).

Accepting Seebohm's argument that the regional ontologies of organic wholes cannot be materially reduced to the ontologies of inorganic wholes, I wish to address another question in the rest of this paper, the question of whether Husserlian mereology can properly describe the whole-parts relations in the system of categories of the hard science of chemistry. The answer to this question is pertinent to the question of whether or not the reduction of chemistry to physics is possible. If it is not possible to reduce chemistry to physics, then this would establish that each regional ontology requires formal ontological theories on the same level of universality.

10.4 The Application of Summative Extensional Mereology to the Lower-Order Regional Ontology of Chemical Wholes

One of the fundamental questions in contemporary philosophy of chemistry is that of the autonomy of chemistry as a science, an issue that is directly related to the reduction of chemistry to physics. In fact, the issue of the reduction of chemistry to physics has been called "one of the main areas in which philosophical interest in chemistry should be directed" (Scerri and McIntyre 1997, 214). Since the ontological dependency of chemical properties on fundamental physical states is not at issue, the sort of reduction meant here is epistemic, rather than ontological, and the question is "whether our current description of chemistry can be reduced to our most fundamental current description of physics, namely quantum mechanics—and with its explanatory consequences" (215). To put this differently, the issue here is not whether the reduction is an ideal possibility since, as Seebohm states, "all entities under the formal ontological categories of the whole and the parts can be considered as units and collections of units" (2015, 243). Thus, a reduction of the material ontological structures of chemical entities is an ideal possibility. However, as with organic entities, the issue is whether such a reduction is a material possibility and this question is what the rest of this essay will address, focusing on the mereology of whole-parts relations for chemical structures.

Despite the received view among many philosophers of science (see, e.g., Hilary Putnam and Paul Oppenheim 1958), quantum chemists and philosophers of chemistry have serious doubts about whether the ontological dependency of chemical states upon physical states undermines the epistemological and explanatory autonomy of chemistry as a science. The hope for such reduction

> seems to have been abandoned and … all that remains is the possibility for approximate reduction. However, criteria for approximate reduction have not been put forward and the notion remains vague … the calculation of the ground state energies of atoms has been achieved to a remarkable degree of accuracy and similarly calculations on small or even medium-sized molecules have given encouraging results. However, whether one can draw the conclusion that chemistry has been reduced rather depends on one's criteria of reduction. If we are to define approximate reduction as has been suggested … then it must be concluded that *chemistry is not even approximately reduced to quantum mechanics*. (Scerri 1994, 168)

As far back as 1939, Linus Pauling claimed that "[a] small part of the body of contributions of quantum mechanics to chemistry has been purely quantum mechanical in character: only in a few cases, for example, have results of direct chemical interest been obtained by the accurate solution of the Schrödinger wave equation … The principal contribution of quantum mechanics to chemistry has been the suggestion of new ideas, such as the resonance of molecules among several electronic structures with an accompanying increase in stability" (1960, vii). However, regarding the chemical concepts of valence and bonding, there is a great deal of doubt about the possibility of reduction.

As well, the conceptual reduction of notions such as composition and molecular structure is also in serious doubt, in part because these concepts do not represent ontological features of the world. When it comes to such concepts, reduction is just not possible even in principle (i.e., it is not even an ideal possibility) "due to the very nature of the concepts themselves. That is, the concepts of composition, bonding, and molecular structure cannot be expressed except at the chemical level … we can calculate certain molecular properties, but we cannot point to something in the mathematical expressions which can be identified with bonding. The concept of chemical bonding seems to be lost in the process of reduction." (Scerri and McIntyre 1997, 218–219). There is also doubt regarding the amount of insight that quantum theory can provide for understanding such chemical concepts. "Many calculations have been extremely sophisticated, designed by some of the foremost researchers in this field to extract a maximum of insight from quantum theory. For simple molecules, outstanding agreement between calculated and measured data has been obtained. Yet, the concept of a chemical bond could not be found anywhere in these calculations. We calculate bonding energies without even knowing what a bond is" (Primas 2013, 5).

Robert Mulliken, who was actively involved in the development of quantum chemistry, claims that "attempts to regard a molecule as consisting of specific atomic or ionic units held together by discrete numbers of bonding electrons or electron-pairs are considered as more or less meaningless, except as approximations in special cases, or as methods of calculation" (1932, 55). For Mulliken, the atom does not exist in a molecule because each orbit is delocalized over all the nuclei and can contribute a stabilizing or destabilizing energy contribution to the total energy of the molecule (see Banchetti-Robino and Llored 2016). "A molecule is here regarded as a set of nuclei, around each of which is grouped an electron configuration closely similar to that of a free atom in an external field, except that the outer parts of the electron configurations surrounding each nucleus usually belong, in part, jointly to two or more *nuclei*" (Mulliken 1932, 55). Thus, there is a key semantic shift here from the concept of molecular *orbit* to that of molecular *orbital*, which takes its meaning from Max Born's probabilistic interpretation according to which the square of a molecular orbital corresponds to the probability density of finding a particular electron within the molecular space. Molecular orbitals have a wave function that contains one electron and that can be delocalized either over all the nuclei or simply over a set of particular nuclei. Thus, the complete electronic wave function is restricted to one of several types that depend upon the symmetry of the nuclear skeleton.

Besides the concepts of molecule and chemical bond, at least two other chemical concepts resist reduction. The first of these is the concept of molecular shape, which cannot be reduced in principle because it is a "mere" concept, albeit one with strong heuristic power. Molecular "shape is metaphorical in virtue of being only chemical ... Molecules can lack an orientation in three-dimensional space, and a particular shape is dependent on the way that the molecule is picked out in measurement. . . . Whether we need to employ the fixed nucleus picture, separate nuclear and electronic 'clouds', or interacting clouds depends on the particular molecule chosen for study, the experimental technique we employ and the questions we ask. As it turns out, there are many different representations of the same property." (Ramsey 1997, 234–248). The second of these other concepts that resist reduction is chemical composition, a concept related to that of bonding and molecular structure. Chemical composition cannot be reduced because "what is physical about a chemical system are its components rather than the system itself [*qua* chemical system] which possesses emergent (though explainable) properties in addition to physical properties" (Bunge 1982, 210). As Jeffry Ramsey states, "the fundamental idea that molecules are constructed additively of atoms, which retain their essential identity within the molecule, is brought into question" (1997, 233). This irreducibility of chemical systems to their physical components is best understood by examining the mereology of systems such as molecules, for example. As Rom Harré and Jean-Pierre Llored explain, "constituent atoms of molecules are not parts of those molecules when we look at the total entity in the light of molecular orbitals. Unlike chair parts which preserve their material properties whether in the chair or on the bench" (2011, 73). It is also questionable whether molecules have definitive physical component parts in such a way that the concept of molecule could be reduced to the concept of physical components without remainder. According to chemist and philosopher of chemistry Joseph Earley, for example, "Na and Cl ions are not parts of salt lattices after that salt has been dissolved. Being in the solution determines that the solution will afford salt as a mass substance on the carrying out of certain operations on sea water, and not something else. Thus, they are at best potential material parts of salt" (2008b, 71).

Harré introduces the concept of "affordance" to capture this idea so that, although there is no salt as such in the sea, the sea "affords" salt under certain conditions and via certain procedures that are carried out on sea water. Philosophers of chemistry argue that, in the same way, there are no atoms in molecules, although "molecules afford atoms in the context of certain manipulations as studies of molecular reactivity have shown us. The material content of a molecule can only be a fusion of atomic potentials, not of atoms [and] affordances are not simple conditional properties … they incorporate the procedure or method used to display their empirical manifestations … the parts of chemical wholes like molecules and atoms are affordances, not themselves concrete entities" (Harré and Llored 2011, 69). This implies that the concept of molecules cannot, therefore, be reduced to the concept of constituent atoms, because these parts of the whole are not concrete entities as such and, in some cases, may be "ephemeral" individuals. For example, "the swiftly composing and decomposing hydrogen-oxygen structures of which real water is really composed are ephemeral individuals. Water is made up of these beings. As such they are [ontological] constituents of a certain whole" (73). But water, as a chemical composition, cannot simply be reduced to these ephemeral parts or occurrences.

As Earley states, "most philosophers have yet to recognize that, when components enter into chemical combination, those components do not, in general, maintain the same identity that they would have absent that combination" (2003, 89). Simple examples that illustrate this point are H_2O and silver chloride. While the property of being H_2O or of being silver chloride "supervenes on the features of the constituent atoms, the features of the atoms on which it supervenes includes features that the atoms have only by virtue of being parts of that compound. The atomic interrelations that give rise to the compound would not obtain if the atoms were parts of a different molecular type" (Francescotti 2007, 58). Molecules themselves are defined in accordance with chemical reaction networks and not vice versa.[6] Since the emergent properties of the whole affect the properties of the constituent parts so that these parts are different than they would otherwise be, the emergent property not only displays novel causal powers, but it very specifically displays downward causal influence. In fact, the causal influence between parts and whole in chemical systems is not asymmetrical. That is, it is neither wholly upwardly directed nor wholly downwardly directed but is, instead, symmetrical in that there is both upward and downward causal influence. As Harré and Llored suggest, this symmetrical causal influence between a chemical whole and its parts is best accounted for through the notion of relationality, because the relations of the parts to one another, as well as the relations of their properties to one another, and the relation of the whole system to its environment are what account for the emergent properties of the whole and its parts. An analysis of relationality in this context, however, requires the development of an adequate mereology that accounts for the fact that, once individuals have entered combinations of interesting sorts, they are no longer the same individuals that existed prior to the combinations.

The classical extensional and summative mereology advocated by Husserl for formal ontology is not adequate for this sort of account because, in its theory of wholes and parts, classical mereology does not consider the relationality of the parts nor does it consider the environment in which parts and wholes exist as something to be factored into the analysis of parts and wholes. In fact, "the standard mereology for chemical compounds involves the presupposition that just as molecules are ultimate constituents or parts of material things, so atom-cores are parts or constituents

[6] I am grateful to Jean-Pierre Llored for suggesting this useful example.

of molecules" (Harré and Llored 2013, 133). According to Harré and Llored, the conceptualization of electrons as entity-like beings involves two mereological fallacies. The first of these is the fallacy of applying to a part a predicate designed to ascribe an attribute to the whole. "A holistic predicate is not necessarily a part predicate, the notion of use is crucial because … predications are context-sensitive" (133). The second mereological fallacy involved in this sort of predication is that of "inferring that substantive products of an analytical procedure are parts of the substance on which the procedure was performed" (134).

Given the discussion above, we note that chemical wholes violate the classical principle discussed by Lewis, which requires that a mereological sum or fusion have no causal effectiveness apart from the causal effectiveness of the parts. As well, quantum chemists consider the classical mereological principle of uniqueness of composition to be "unintuitive and inadequate to those rules for chemical parts-whole reasoning that are required to accommodate the role of chemical entities in structures, such as atoms in 'polyatomic' ions. [Quantum chemists] are likely to question the relevance of transitive mereology whenever the whole molecule and the parts are co-dependent and relative to a specific environment or action" (Llored 2014, 159). Regarding the classical principle of standard mereology discussed by Fine, that different methods of composition do not yield different wholes, we must point out that Fine himself admits that mereology must recognize relationships, as well as other sorts of parts, in a way that does not conflate relations between parts with properties of parts. This is important, for example, in order to accommodate the mereology of structured wholes, such as chemical molecules, since these "generally have causal efficacy in virtue of their 'connectivity'—in addition to the causal powers of their constituent atoms, levorotatory amino acids are nutritious [but] the corresponding dextrorotatory amino acids are poisonous—although both sorts of molecules have exactly the same component parts" (Earley 2008a, 9). However, in addition to optical activity and chirality, we find that the structural arrangement of the parts of chemical wholes also contributes to the properties of the whole. Thus, different structural arrangements of the same parts will yield different chemical wholes. This characteristic of molecules, which is known as structural isomerism, was first discovered in 1827 by Friedrich Woehler, when he prepared cyanic acid and noted that although its elemental composition was identical to fulminic acid, its properties were quite different. This discovery challenged the prevailing chemical understanding of the time, which embraced the classical mereological principle that chemical compounds could be different only when they had different elemental compositions.

It is, therefore, clear that the properties of chemical wholes are not merely a function of the properties of their parts. As well, once we examine closely the way quantum chemists conceive of molecules, atoms, and subatomic "parts," we notice that they think of these in terms of chemical relations and relata at different levels of organization that co-define each other. What is needed then is a non-standard mereology that captures this co-defining and co-constituting feature of chemical whole-parts relations. To the extent that parts and whole co-constitute each other and to the extent that parts and whole are co-dependent, the parts must be considered as moments of the whole rather than as pieces.

But there is more. Chemical mereology must also capture the way the environment (or milieu) of the chemical experiment contributes to constituting chemical whole-parts relations. Such a mereology challenges the assumptions of classical summative mereology since, in quantum chemistry, it is clear that the parts do not define the whole but, rather, the parts and the whole co-define each other in the context of chemical activity within a given environmental context. Yet, for such a mereology to succeed, our understanding of chemical parts and wholes as independent concepts must itself be altered and, to do this, we turn to the way in which quantum chemists involve different levels of organization in their daily experimental practice, in their syntheses and analyses of chemical structures, and in their calculations.

Developments in quantum chemistry have shown that the constituents of atoms and molecules should not be described as entity-like or as having clear boundaries in the way in which atoms or molecules were traditionally conceived as bounded and entity-like (e.g., ball and stick models). In fact, as Harré and Llored emphasize, "it is a mistake to treat electrons as constituents of anything" (2013, 159). Thus, the classical summative mereology advocated by Husserl in the context of formal ontology is not only inadequate for describing whole-parts relations for organic wholes as Seebohm argues, it is also inadequate for describing whole-parts relations for the regional ontology of quantum chemistry, which deals with "such structured entities as molecules, where the whole is not merely the sum of its parts" (Sukumar 2013, 303).

At best, then, "the part-whole relation … may behave according to such principles as [discussed by Husserl]. But there is a growing consensus that this is the best one can say, and that mereology is best understood as a theory—or a plurality of theories—whose fundamental truths do not reflect the properties of the part-whole relation itself but the nature of the entities to which it applies" (Seebohm 2015, 16). If this is the case then, the analysis of the part-whole relation with regard to both the higher-order ontologies of organic wholes and the lower-order ontology of chemical wholes serves to illustrate the inadequacy of formal mereology for regional ontologies, thus calling into question Husserl's belief that mereology is an element of formal ontology. As well, these discussions draw attention to the need for developing non-summative and non-extensional mereologies that are tailored to describe whole-parts relations in the various regional ontologies studied by the natural, human, and social sciences.

References

Banchetti-Robino, M.P., and J.P. Llored. 2016. Reality without reification: Philosophy of chemistry's contribution to philosophy of mind. In *Essays in the philosophy of chemistry*, ed. Eric Scerri and Grant Fisher, 83–110. Oxford: Oxford University Press.

Berque, A. 2010. *Milieu et identité humaine. Notes pour le dépassement de la modernité*. Paris: Donner lieu.

Bunge, M. 1982. Is chemistry a branch of physics? *Journal for General Philosophy of Science* 13: 209–223.

Earley, J.E. 2003. Varieties of properties: An alternative distinction among qualities. *Annals of the New York Academy of Science* 988: 80–89.
———. 2008a. How philosophy of mind needs philosophy of chemistry. *HYLE – International Journal for the Philosophy of Chemistry* 14: 1–26.
———. 2008b. How chemistry shifts horizons: Element, substance and the essential. *Foundations of Chemistry* 11: 65–77.
Fine, K. 1994. Compounds and aggregates. *Noûs* 28: 137–158.
Francescotti, R.M. 2007. Emergence. *Erkenntnis* 67: 47–63.
Gruszczyński, R., and A. Varzi. 2015. Mereology then and now. *Logic and Logical Philosophy* 24: 409–427.
Harré, R., and J.-P. Llored. 2011. Mereologies as the grammars of chemical discourses. *Foundations of Chemistry* 13: 63–76.
———. 2013. Molecules and mereology. *Foundations of Chemistry* 15: 127–144.
Husserl, Edmund. 1970. *Logical Investigations*, 2 vols. Trans. J. N. Findlay. New York: Routledge & Kegan Paul.
Lewis, D. 1986. Against structural universals. *Australasian Journal of Philosophy* 64: 25–46.
Llored, J.-P. 2014. Whole-parts strategies in quantum chemistry: Some philosophical and mereological lessons. *HYLE – International Journal for the Philosophy of Chemistry* 20: 141–163.
Mulliken, R.S. 1932. Electronic structures of polyatomic molecules and valence I. *Physical Review* 40: 55–62.
Pauling, L. 1960. *The nature of the chemical bond and the structure of molecules and crystals*. 3rd ed. Ithaca: Cornell University Press.
Primas, H. 2013. *Chemistry, quantum mechanics and reductionism: Perspectives in theoretical chemistry*. Berlin: Springer.
Putnam, H., and P. Oppenheim. 1958. Unity of science as a working hypothesis. In *Minnesota studies in the philosophy of science*, ed. H. Feigl, M. Scriven, and G. Maxwell, vol. II. Minneapolis: University of Minnesota Press.
Ramsey, J. 1997. Molecular shape, reduction, explanation and approximate concept. *Synthese* 111: 233–251.
Scerri, E.R. 1994. Has chemistry been at least approximately reduced to quantum mechanics? *PSA: Proceedings of the Biennial Meeting of the Philosophy of Science Association* 1994: 160–170.
Scerri, E.R., and L. McIntyre. 1997. The case for the philosophy of chemistry. *Synthese* 111: 213–232.
Seebohm, T.M. 2015. *History as a science and the system of the sciences: Phenomenological investigations*. New York: Springer.
Simons, P.M. 1982. Three essays in formal ontology. In *Parts and moments: Studies in logic and formal ontology*, Analytica series, ed. Barry Smith, 111–260. Munich: Philosophia Verlag.
Smith, B., and K. Mulligan. 1982. Pieces of a theory. In *Parts and moments: Studies in logic and formal ontology*, Analytica series, ed. Barry Smith, 15–110. Munich: Philosophia Verlag.
Sukumar, N. 2013. The atom in a molecule as a mereological construct in chemistry. *Foundations of Chemistry* 15: 303–309.

Marina Paola Banchetti-Robino received her Ph.D. from the University of Miami. She is currently an Associate Professor at Florida Atlantic University, where she is also serving as Chair of the Department of Philosophy. She specializes in phenomenology, philosophy of language, philosophy of mind, and philosophy of science with a focus on history and philosophy of chemistry. She has published articles in *Synthese*, *Husserl Studies*, *Philosophy East and West*, *Continental Philosophy Review*, *The Review of Metaphysics*, and *Foundations of Chemistry*. She has also co-edited two books, *The Philosophies of Environment and Technology* (1999) and *Shifting the Geography of Reason* (2007).

Chapter 11
Science, Intentionality, Control, and the Strata of Experience

Harry P. Reeder

Abstract This chapter uses a comparison between Seebohm and the social scientist Felix Kaufmann as a way of introducing Kaufmann, a fascinating figure in his own right who was both of a member of the Vienna Circle and a phenomenologist to a wider philosophical audience, and to locate Seebohm's positions within the broader framework of discussions in the philosophy of the sciences that were so central to both the analytical tradition and—at least in its inception—the phenomenological tradition as well.

This essay attempts to offer a brief summary of some important elements of Felix Kaufmann's many discussions of the methodology of the sciences,[1] and to indicate some important similarities and differences between them and Thomas M. Seebohm's recent work, *History as a Science and the System of the Sciences: Phenomenological Investigations* (2015).[2]

Felix Kaufmann was a permanent member of the Vienna Circle, but he was never a logical positivist, or, as the positivists came to call themselves logical empiricists (1939b, 434). He was also a member of a research group working with Ludwig von Mises, one with Hans Kelsen, and a group calling themselves "Geist-Kreis" (Helling 1984, 1985). He was a close follower of Edmund Husserl's phenomenology (1968, 138), although he understood the positivists' views very well (1940, 1950). In meetings of the Vienna Circle he often referred to himself as "His Majesty's opposition" (1939b, 436) and wrote that he learned much from his discussions with members of the Vienna Circle.[3] He rejects the physicalism of Carnap and Neurath, praising

[1] A more complete bibliography of Kaufmann's published works may be found in Reeder (1991).

[2] All references to Seebohm are to this work.

[3] Letter to Arthur Bentley dated February 15, 1945. This letter is NP 15467. In the present discussion 'NP' refers to pages in Kaufmann's *Nachlaß*, which is located in the archives of the Center for Advanced Research in Phenomenology, at the University of Memphis. A microfilm of the *Nachlaß* is located at the University of Konstanz, Germany.

H. P. Reeder (deceased)
University of Texas at Arlington, Arlington, TX, USA

© Springer Nature Switzerland AG 2020

T. Nenon (ed.), *Thomas Seebohm on the Foundations of the Sciences*, Contributions to Phenomenology 105,
https://doi.org/10.1007/978-3-030-23661-8_11

Husserl's *Formal and Transcendental Logic* and *Experience and Judgment,* saying that in those works:

> The task of laying bare the strata in the constitution of the objective world is attacked with previously unattained profoundness. The exposition of this problem might well form the best means of access for logical empiricists to the fundamental concept of transcendental phenomenology, the concept, namely, of "*epoché*." (1968, 138)

Kaufmann frequently refers to himself a *methodologist*, because his main interest is to analyze the nature of scientific method. Likewise, Seebohm focuses upon the issue of methodology on the very first page (2015, 1) of his Introduction, and seeks to offer analyses that contribute to a "phenomenological answer to the general question 'what are the empirical sciences' and to offer some guidelines for further investigations in the different branches of a phenomenological epistemology" (7).

Kaufmann uses the phrase "strata of experience" to refer to what Husserl calls the levels of constitution of meaning, including that of the objective world (1940–1941, 324; 1968, 138). He uses this notion to indicate the hierarchical nature of the rules of scientific procedure. Nonetheless, in the Prologue to *Methodenlehre der Sozialwissenschaften* (1936a, iii–iv; Spanish translation *Metodología de las ciencias sociales* [1946]) he distances his methodology from a purely transcendental phenomenology, referring to the work as a "formal critique" rather than a purely phenomenological "transcendental critique." Nonetheless Kaufmann introduced Alfred Schutz to Husserl's phenomenology and together they made careful studies of Husserl's *Logical Investigations, Ideas Pertaining to a Pure Phenomenology and to a Phenomenological Philosophy*, *Formal and Transcendental Logic*, *Cartesian Meditations*, *On the Phenomenology of the Consciousness of Internal Time*, and drafts of *Experience and Judgment*. Kaufmann went on to introduce Schutz to Husserl personally, and Husserl invited Schutz to publish reviews of some of his works. In his acknowledgements Seebohm singles out Kaufmann and Schutz (among others) as ineliminable precursors to his book (2015, ix).

As a methodologist Kaufmann seeks to clarify the nature of scientific method, in both the natural and the social sciences, with a special interest in the social sciences (1936a; 1944). As a matter of emphasis, Seebohm expands the focus upon the social sciences to include (or open the way to) discussions of, e.g., actions, speeches, texts, monuments and artifacts (2015, vi), religion and art (6).[4] For instance, Seebohm's broader view may be seen in the following statement: "A phenomenological epistemology in the broad sense as a general theory of knowing is interested in the different qualities, scopes, limits and temporal dimensions of the *evidence* in which the objective correlates of intentionality are given" (40–41). In what follows we will sketch Kaufmann's general methodological views of the sciences. Perhaps at the start it is important to stress that—unlike many former and contemporary philosophers—Kaufmann insisted that the differences between the natural and the social

[4] Seebohm's focus in chapters 2–6 are clearly broader than Kaufmann's discussions.

sciences are differences of *degree* rather than of *kind* (1938, 452, 456, 459–461; 1944 169–170, 177). Various features of this view are clarified below.[5]

For Kaufmann the sciences are defined not by a corpus of propositions, nor by deductive logic alone, although he tries to clarify the relation between these and scientific method. Kaufmann holds that the sciences are to be defined in terms of the *rules of scientific procedure* (1944, 42). Between his *Methodenlehre der Sozialwissenschaften* (1936a) and his *Methodology of the Social Sciences* (1944), his view of methodology underwent a change. He attributes this change to his study of John Dewey's *Logic, the Theory of Inquiry* (1938). This change is in his understanding of the relationship between deductive logic and methodology: "I came to the conclusion that methodology must be clearly distinguished from deductive logic and recognized as an autonomous rational discipline" (1944, vii), a discipline that is separate from, but using many insights from, Husserl's transcendental phenomenology. This change in his view stemmed largely from his identification of one of the most important basic rules of scientific procedure: the principle of permanent control. Citing Einstein's famous observation that, "In so far as the axioms of mathematics refer to reality they are not certain; and in so far as they are certain, they do not refer to reality,"[6] Kaufmann insists that, because any empirical proposition may be overturned by new empirical evidence (which he called the Principle of Permanent Control), the a priori rules of deductive logic may not be used to isolate any empirical claim from possible rejection based on future empirical evidence, although they may be used as "conventions" or "heuristic postulates" to isolate the empirical application of the rules of deductive logic and other basic rules of procedure from over-hasty rejection of future empirical claims. Seebohm agrees: "Even a theorem in the natural sciences that has been confirmed up until now in experimental research can be disconfirmed by further experiments in the future"[7] (2015, 35 n. 66). To indicate the roles of key logical laws in scientific procedure, Kaufmann introduces the basic rules of the "procedural correlates" of the principles of contradiction and of the law of the excluded middle (1944, 231).

Kaufmann's notion of "heuristic postulates" is a significant addition to the epistemology of the sciences, which separates his view of scientific method from both the "apriorist" claim (as he referred to it) that deductive logic is the basis of all scientific method (e.g., Karl Popper) and the radical empiricist claim that only inductive logic can serve as such a basis:

> The "heuristic postulates" which tell us to apply a particular method are, in so far as they involve the thesis that it is the only right method, also conventions. These assertions, too,

---

[5] For a fuller discussion of these and related topics see Reeder (1991, chaps. 1–3).

[6] Einstein's *Geometrie und Erfahrung*, Berlin, 1921, p. 3, as cited in Kaufmann's (1938, 456). Einstein considered Kaufmann to be "one of the great philosophers of our time" (related in a letter from Liesel Marmorek dated August 7, 1948; NP 10687, and reported to have been revealed by Einstein's secretary).

[7] It is important to note that, because of the notion of heuristic postulates, "disconfirmed" here is quite different from Popper's refutation by one instance of empirical evidence because of the logical law of modus tollens.

may be, and often are, supported by experience, but their conventional character rests on the fact that the possibility of their refutation by experience is suspended. (1934, 108)

The rules of the methodology of a particular science "define the *structure* of [that] science. To this corresponds a certain *corpus of knowledge* varying in time" (1940–1941, 317). The phrase "varying in time" here refers in part to the principle of permanent control. According to Kaufmann "the nucleus of the logic of empirical science (methodology) is the analysis of empirical procedure determining the criteria of correct scientific decisions in given situations" (1945, 469). A *scientific decision* is the decision to admit or reject a proposition from the scientific corpus in a given scientific situation:

By *scientific procedure* we decide whether given synthetic propositions should be considered parts of the corpus of a given science. Decisions to change the status of a proposition in this respect are called *scientific decisions*. Unambiguous meanings of the sentences concerned are presupposed in scientific decisions. (1944, 229; emphasis added)

The presupposition of unambiguous meanings here further reflects the principle of permanent control, as one of the main goals of methodology is to clarify and remove ambiguities of meaning, as far as is possible in a given scientific situation. The rules of scientific procedure cannot be determined by deductive logic alone:

Failure to distinguish between deductive logic in the strict sense and the logic of scientific procedure (methodology) leads to an inadequate (elliptical) formulation of scientific problems, methods, and solutions. Replacing the elliptical by complete formulations, we see that it is possible to prove the correctness of scientific decisions *in terms* of the rules *on the basis* of established knowledge. In other words, 'Correct scientific decision' is defined in terms of the rules. *The logic of science (methodology) is the theory of correct scientific decisions*. (1944, 230)

Note that Kaufmann's methodological focus here is on scientific method rather than on human experience in general. This is perhaps the main reason that he separates methodology from Husserl's transcendental phenomenology, although he certainly incorporates many phenomenological insights within his scientific methodology. Central to his use of phenomenological descriptions is his insistence on the roles of intersubjectivity, interpretation, implicit theories of perception, and his rejection of what he (with Husserl) refers to as "sensualism." Kaufmann doesn't use the Husserlian term "intersubjective," but Kaufmann's concept of the principle of permanent control includes the sense of Husserl's intersubjectivity. For Kaufmann sensory evidence, instead of presenting "the bare facts," always contains interpretation and implicit theories, evidence that is subject to intersubjective control, largely by language (see below). Thus sensory evidence must be intersubjective, or, as Seebohm suggests with a useful new term, it must be "intersensory" (2015, v). However, before entering into these discussions it is useful to outline some results of Kaufmann's discussions of scientific method, in order to guide the discussions.

Most of Kaufmann's publications attempt to clarify the ambiguities in differing views in many philosophical controversies concerning the nature of scientific thought, especially in the social sciences, showing that they are often based in faulty views of the relation between logical and empirical elements of scientific method (cf. Seebohm 2015, v–vi).

Kaufmann identifies the two most basic strata of the rules of scientific procedure, which he calls *basic rules* and *preference rules*: the *basic rules* of a science provide the criteria for the truth or falsity of propositions (1939a, 541). The *preference rules* of a science presuppose the basic rules:

> Frequently a choice will be made between alternative hypotheses of methods without any reason being given; but it is a tendency of scientific thought to reduce unexplained decisions to a minimum.… Rules which determine the choice to be made are what we call "*preference rules*." (541; emphasis added)

Elsewhere he makes the same distinction by dividing the rules of scientific method into two types: rules of language (basic rules that apply only to meanings of scientific terms but not to the empirical world) and rules of procedure (which apply to empirical investigations) (1940, 65; cf. 1940–1941, 315). According to Kaufmann, blurring this distinction is an important cause of many controversies over the nature of scientific method. Otherwise put: the basic rules of procedure "define 'correct scientific decision in a given situation,' without reference to the goals of inquiry," whereas the preferences of procedure are relative to goals of inquiry, such as "presumable relevance," and "the degree of significance of given problems in terms of theoretical ideals, such as unity, simplicity, universality, and precision of laws" (1944, 232). One must recall here that insofar as procedural rules are applied to the empirical world even the basic or presupposed meanings of terms are subject to the principle of permanent control.

The difference between the two types of rules indicate the relative status of them: basic rules are more fundamental and are presupposed in all scientific situations, for without them it would not be possible to recognize the unity of a particular science as the *same* science (1940–1941, 317–318). However the basic rules are not a priori in the sense of the laws of deductive logic, since even they are subject to the principle of permanent control (for instance, consider the long life of the meaning of the term "atom" in classical physics, the discovery of sub-atomic particles, and the radical change of the meaning of the term with the advent of quantum physics). The basic rules require quite a bit more empirical evidence to be called into question. Thus they are presupposed a priori "in the sense that—again because they make no assertions about reality—they cannot be refuted by 'experience' (observation); and they are a priori for science because 'science' is defined in terms of them." But the basic rules are not a priori in the sense that they are "irrefutable statements about reality" or "being derivable (as are the rules of deductive reasoning) from the propositions to which they apply" (1939a, 46).

Kaufmann occasionally indicates the commonality of scientific and ordinary experience: "The basic elements of empirical procedure are common to pre-scientific and scientific thinking, and there is indeed no sharp line of demarcation between them."[8] Seebohm agrees: "… the objective validity of knowledge is only

[8] Note that this is in line with Husserl's *Crisis*, in which he examines at length the embedding of scientific thought in the pre-scientific thought of the lifeworld.

possible as intersubjective validity and … sensory observation can, therefore, serve as objectively valid observations only as intersensory, i.e. intersubjective sensory, observation" (2015, 3–4). Note here a key role of language in interpreting intersensory observation. Nonetheless, Kaufmann's focus is almost always upon scientific method, with relatively rare references to broader issues of the life world.[9] One can thus say that Seebohm's "phenomenological epistemology" is, in general, broader than Kaufmann's focus upon science. Nonetheless, in his introduction Seebohm says that: "The main intention of the following investigation is, hence, to give a phenomenological answer to the general question 'what are the empirical sciences' and to offer some guidelines for further investigations in the different branches of a phenomenological epistemology" (2015, 7). The latter part of this quotation indicates that his interest in fact is wider than might be expected on the basis of his title: *History as a Science and the System of the Sciences: Phenomenological Investigations*.

In various publications Kaufmann provides several lists of the elements of scientific method. These various lists may be summarized as follows:

1. demarcate (separate) sciences from one another and from ordinary experience;
2. make their presuppositions explicit and coherent, so that they may be critiqued intersubjectively;
3. identify rules for making warranted and unwarranted scientific decisions;
4. make the canons of scientific procedure explicit;
5. show the relations between observations and other types of scientific control.[10]

The fifth rule, above, is further clarified by Kaufmann in "The Significance of Methodology for the Social Sciences" as follows:

> … no statement about man is based merely on external data, but always includes an *interpretation* of these data on the basis of self-observation (introspection). There is thus a relationship between data of external experience and data of self-observation, which admits of mutual control between the two types of data … the psychic bases of explanation cannot be excluded from the sciences of human behavior … It follows that the "understanding" of human behavior is not coordinate with the explanation of natural science, but that, since it is the more complex process, it includes the latter. (1938, 452–453, emphasis added)

He clarifies this "mutual control" elsewhere: "The interpretation of meaning is always a synthesis of internal and external experience, and accordingly concepts of social facts must be constructed out of concepts of physical and of psycho-physical facts" (1936b, 78). These psycho-physical facts include that of man as an inhabitant of a spatio-temporal world (1940–1941, 323).[11]

---

[9] A notable exception is "Strata of Experience" (1940–1941).

[10] Reeder (1991, 11)—this list is a summary of various passage from Kaufmann's works, such as (1944, 39–40) and (1943–1944, 268–269). On the notion of intersubjective critique, see Seebohm (2015, 40–41).

[11] This is Kaufmann's manner of referring to what Husserl calls in the *Crisis* the intersubjective constitution of the life-world.

Kaufmann also expresses this notion of "interpretation" in terms of implicit theories of experience, and one of the key features of his methodology of the sciences is to make these implicit theories explicit, to permit their analysis and critique:

> It is impossible to isolate perceptual content from principles of form. Every perceptual experience involves processes of selection in conformity with certain general patterns. This means that theory is implicit even in the most "simple" experiences. The view that assertions are tested by confrontation with "pure sensory experience" is therefore an inadequate interpretation of the observational test.[12] (1945, 471)

Here one can see Kaufmann's major critique of positivism: its rejection of the relevance of introspection to scientific thought. Kaufmann expresses this more forcefully in "Phenomenology and Logical Empiricism," where the connection between methodology and phenomenology is also clarified, showing his interest in the methodological investigation of both scientific and non-scientific experience:

> He who recognizes the highly complex structure of the psychic acts by which objects are "given" also comprehends the significance of an analysis of these acts. Thus he gains initial access to the problems of phenomenology and an understanding of one of its central concepts, the concept of intentionality. (1968, 132)
>
> ... *evidence denotes the intentional accomplishment belonging to the giving of "it itself" (der Selbstgebung).* (136)[13]

Seebohm places intentional analysis squarely in the heart of a phenomenological epistemology: "A general theory of knowledge appears in the context of phenomenological investigations as the intentional analysis of the relation between intentional objects or *cogitata* and intentional acts or *cogitationes*" (2015, 2). Again, note here that Seebohm is interested in opening up a general phenomenological theory of knowledge, which is broader than Kaufmann's focus upon science. A further example of Seebohm's wider view can be seen in his Preface:

> What has to be explained [in the human sciences] are manifestations of cultural activities, i.e., actions, interactions, speeches, but then also written speeches, texts, art works, etc. Such manifestations are more than objects that be given in intersensory observations. They must be understood, i.e., they need interpretations. Explanations in the human or cultural sciences presuppose, hence, interpretations. There is no way to defend the objectivity validity of such explanations without a possible justification of the objective validity of the presupposed interpretations with the aid of methods that can serve as warrants for the objective validity of these interpretations. (vi)

---

[12] Cf. Husserl, "Philosophy as a Rigorous Science" (1965), especially p. 73: "As far as science, real science, extends, so far can one teach and learn, and this everywhere in the same sense. Certainly scientific learning is nowhere a passive acceptance of matter alien to the mind. In all cases it is based on self-activity, on an inner reproduction, in their relationships as grounds and consequences, of the rational insights gained by creative spirits." It is instructive to note two things here: first, the reference to teaching and learning necessarily requires the role of language in scientific procedure (and in a phenomenological epistemology), and, secondly, the reference to creative spirits necessarily requires the notion of science as an intersubjective testing and agreement upon scientific *procedure*.

[13] Cf. Husserl (1965 89): "... if knowledge theory will nevertheless investigate the problems of the relationship between consciousness and being, it can have before its eyes only being as the correlate of conscious of consciousness, something 'intended' after the manner of consciousness ..."

Seebohm's interest in a *general* theory of knowledge, rather than merely a phenomenological analysis of scientific knowledge, is expressed as follows: "A general theory of knowledge covers the entire realm of intentional acts and their correlates, the intentional objects, and their foundations in passive syntheses" (3). His discussions of history as a science (part II) expand the notion of "science" beyond the usual discussions of the "hard" sciences (physics, chemistry, biology, etc.). Seebohm's analysis of history is much more specific than Kaufmann's many references to the "social sciences." As such, Seebohm's analyses can be seen to be based in part upon Kaufmann's general insights but pursues some of them to a greater depth.

Kaufmann's focus, on the contrary, is upon the rules of scientific procedure. In "Strata of Experience" Kaufmann distinguishes between what might be called the elements of experience that are relevant to science and those that are irrelevant. The *givenness* of the rules of scientific procedure:

> … does not mean anything else for the methodologist than that the correctness of procedural decisions in this science is defined in terms of these rules … Psychological, sociological, or historical aspects of givenness are by no means excluded, but no methodological conclusions can be derived from them.[14] (1940–1941, 319)

He refers to the positivists' error using the same term that Husserl used in his critique of empiricist philosophy's error: sensualism (1940–1941, 321–322). Kaufmann criticizes the physicalist doctrine of the positivists as being an over-use of Occam's razor (1939b, 436). He counters the physicalistic doctrine of sensualism in yet another way:

> It can be said that all empirical knowledge is relative in so far as it is possible that it may be refuted by later experience, and subjective in so far as it does not consist merely in a passive reception of objectively given "things in themselves" but necessarily implies syntheses of the "material of experience," which latter cannot be thought of in isolation. (1936b, 69)
>
> The cure for the error of sensualism is to be found in careful phenomenological description:
>
> The phenomenological *epoché* does not only preclude all references to assumptions about matters of fact but even all references to already constituted meanings—including the concept of man as an inhabitant of a spatio-temporal world. A step by step reconstitution of meanings in their proper order is posed as the task for the phenomenologist. (1940–1941, 323)[15]

The reference here to "man as an inhabitant of a spatio-temporal world" is related to the distinction above between psychic acts that are relevant to scientific method and those that are irrelevant. The spatio-temporal world is objective in the phenomenological sense, according to which "objectivity" has the sense of "intersubjectively established meanings." It is through the mutually controlled (intersubjective)

[14] Here there is a parallel with Aristotle's formal and final causes. For Kaufmann, as for all modern science, the final (psychological) cause is not methodologically controlled and is thus rejected as part of the methodological rules of science, whereas the formal cause is methodologically (intersubjectively) controlled and is thus central to scientific method.

[15] Husserl seeks to provide the constructions of pre-scientific meanings in *Crisis*, *Formal and Transcendental Logic*, and *Experience and Judgment.*

coherence experience—a control based as much in language as in behavior—that the lived intentional meanings are established. Nonetheless, Kaufmann distinguishes this sort of phenomenological coherence from any sort of coherence theory of truth (1940–1941, 315). Seebohm identifies language as a key element in phenomenological method:

> The minimum requirement for phenomenology as a rigorous science is that it is possible to present such explications and analyses to co-subjects who share the methodical approach of the phenomenological attitude. Science is an intersubjective enterprise. The problem of solipsism will vanish if it is possible to give a phenomenological explication of the givenness of other persons, in short Others, and intersubjectivity. (2015, 36)[16]

In *The Methodology of the Social Sciences* Kaufmann lists five "fundamental properties of the system of rules that are regarded as essential for scientific procedure in general" (1944, 230). They are:

(a) The grounds of a scientific decision must be among the propositions belonging to the scientific situation to which the decision is related.
(b) Observational reports (protocol propositions) play a key-role among the grounds.
(c) All scientific decisions are reversible (principle of permanent control).
(d) A scientific decision must not lead to a scientific situation containing two incompatible propositions (procedural correlate of the principle of contradiction).
(e) No proposition can be in principle undecidable (procedural correlate of the principle of the excluded middle). (1944, 231)

Likewise, in the same work Kaufmann summarizes five genuine methodological questions:

(a) What is the logical form of laws?
(b) Under what conditions are such propositions accepted?
(c) Under what conditions are such propositions eliminated?
(d) How do propositions function as controls for other propositions?
(e) How far does the system of laws established at present fulfill the conditions implied in the ideal of a rational cosmos? (1944, 77)

Taken together, these two lists provide an elegant summary of Kaufmann's methodology of the sciences. He uses them in his examinations of various scientific and philosophical views of science, and one of his main goals is to clarify implicit presuppositions that provide the basis for the many methodological controversies concerning the nature of science and scientific method.

Kaufmann's various formulations of basic laws of science are not in conflict but are complementary (for example the fifth methodological question above includes

[16] The phrase "share the methodical approach of the phenomenological attitude" is important here, as many present phenomenologists deny that there is such a methodical phenomenological attitude. For further discussion of the role(s) of language in the method of phenomenology, see Reeder (2010 Chapter VII and Appendix).

the notions of the scientific situation and the principle of permanent control are included, due to the phrase "at present"). He formulates these basic elements of the rules of scientific procedure in various ways to clarify the errors of both sensualistic empiricist views and what he calls rationalist apriorism, which latter dogmatically asserts that various laws, evidences or truths must simply be accepted as formulated by various philosophers (1940–1941, 321).

What can we say by way of summarizing our findings in this essay? (1) Kaufmann provides a stratified and intersubjective analysis of scientific method in terms of the human procedures necessary to make an endeavor a science.[17] (2) These stratified procedures have a hierarchical structure including both logical and empirical elements, elements that must be carefully distinguished. (3) The fundamental layer of this hierarchy consists of basic rules, rules concerned primarily with the meaning of basic terms, although still subject to the principle of permanent control due to their conventional status. These rules are *a priori* in the sense that they are, at any given time, analytical and non-empirical. (4) Upon these basic rules are founded preference rules that guide the application of the basic rules to empirical investigation. It is in their application to nature based upon preference rules that conventional rules face the principle of permanent control. (5) Every proposition in the *corpus* of a science, at any given time, is subject to the principle of permanent control—even the propositions stating the basic laws of a science. According to this principle every scientific proposition—including those stating basic and preference rules—may be overturned due to future experience and discovery. (6) Kaufmann's focus is almost always upon the nature of scientific method, with relatively rare mention of broader issues concerning the life world. (7) Seebohm's focus is, in general, broader, seeking to embed questions of scientific method within the broader questions broached by a general "phenomenological epistemology." (8) There are frequent overlaps of the investigations of the two thinkers. (9) Both thinkers view their respective investigations in terms of the phenomenological method, although Kaufmann does not see his discussions as stemming solely from the phenomenological method.

References

Dewey, John. 1938. *Logic, the theory of inquiry*. New York: Henry Holt and Company.

Helling, I.K. 1984. A. Schutz and F. Kaufmann: Sociology between science and interpretation. *Human Studies* 7: 141–161.

———. 1985. Logischer Positivismus und Phänomenologie: Felix Kaufmanns Methodologie der Sozialwissensschaften. In *Philosophie Wissensschaft, Aufklärung: Beiträge zur Geschichte und Wirkung des Wiener Kreises*, ed. Hans-Joachim, 237–256. New York: Walter de Gruyter.

Husserl, Edmund. 1965. Philosophy as a rigorous science. In *Phenomenology and the crisis of philosophy*, ed. and trans. Quentin Lauer, 71–147. New York: Harper Torchbook.

[17] Here the term "human" indicates the non-logical and non-strictly-empirical elements of scientific procedure, such as heuristic postulates. For Kaufmann, as for Husserl and Seebohm, science is above all a human, rather than merely a logical, endeavor.

Kaufmann, Felix. 1934. The concept of law in economic science. *Review of Economic Studies* 1: 102–109.
———. 1936a. *Methodenlehre der Sozialwissenschaften*. Vienna: Verlag von Julius Springer.
———. 1936b. Remarks on the methodology of the social sciences. *The Sociological Review* 28: 64–84.
———. 1938. The significance of methodology for the social sciences. *Social Research* 5: 442–463.
———. 1939a. The significance of methodology for the social sciences (Part II). *Social Research* 6: 537–555.
———. 1939b. Unified science (Note). *Social Research* 6: 433–437.
———. 1940. Truth and logic. *Philosophy and Phenomenological Research* 1: 59–69.
———. 1940–1941. Strata of experience. *Philosophy and Phenomenological Research* 1: 313–324.
———. 1943–1944. Verification, meaning and truth. *Philosophy and Phenomenological Research* 4: 267–283.
———. 1944. *Methodology of the social sciences*. New York: Oxford University Press. [Not a translation of *Methodenlehre der Sozialwissenschaften*, 1936.] Reprint, New York: 1958.
———. 1945. The nature of scientific method. *Social Research* 12: 464–480.
———. 1946. *Metodología de las ciencias sociales*. Mexico City: Fondo de Cultura Económica. [Spanish translation by E. Imoz of *Methodenlehre der Sozialwissenschaften*, with a 2 page note at the end dated New York, February 1946.]
———. 1950. Basic issues in logical positivism. In *Philosophic thought in France and the United States: Essays representing major trends in contemporary French and American philosophy*, ed. M. Farber, 565–588. Albany: SUNY Press.
———. 1968. Phenomenology and logical empiricism. In Philosophical essays in memory of Edmund Husserl, ed. M. Farber, trans. Dorion Cairns, 124–142. New York: Greenwood Press.
Reeder, Harry P. 1991. *The work of Felix Kaufmann*. Current continental research series no. 220. New York: University Press of America and the Centre for Advanced Research in Phenomenology.
———. 2010. *The theory and practice of Husserl's Phenomenology*, 2nd enlarged ed., Pathways in phenomenology, vol. II. Bucharest: Zeta Books.
Seebohm, Thomas M. 2015. *History as a science and the system of the sciences: Phenomenological investigations*. New York: Springer.

Harry P. Reeder (Ph.D. Waterloo, 1977) taught at the University of Guelph, Wilfrid Laurier University, and the University of Alberta before joining the department and becoming Professor of Philosophy at the University of Texas at Arlington, where he won multiple teaching awards. His research interests are in the areas of phenomenology, Wittgenstein, the philosophy of language, philosophy of science, critical thinking, and hermeneutics. A few of his many monographs in Spanish and English are *The Theory and Practice of Husserl's Phenomenology* (Bucharest 2010) (translated as *La praxis fenomenológica de Edmund Husserl* (Bogotá 2011)), *La globalización y la genomenología dell lenguaje* (Morelia, Mexico 2009), *Arguing with Care: Dialectic for a Democratic Society* (Dubuque 2009), *The Work of Felix Kaufmann* (New York 1991), and *Language and Experience: Descriptions of Living Language in Husserl and Wittgenstein* (New York 1984). He is currently living and working in Medellin, Colombia.

Chapter 12
Husserl on the Human Sciences in *Ideen II*

Thomas M. Seebohm

Abstract In this chapter, Seebohm situates the projects of the Ideas II and the Ideas III within the context of the 19th century discussion about the nature and grounding of the human sciences as exemplified above all by Mill and Dilthey. The chapter then provides a summary and critical analysis of Husserl's phenomenological grounding of the "spiritual (*geistige*)" sciences, with an aim to demonstrating how it was able to justify Dilthey's main intuitions about the differences between the natural and the spiritual or human sciences and eliminate the shortcomings of Dilthey's approach to the empirical human sciences.

12.1 Introduction

To translate *Geisteswissenschaften* with "human sciences" was a good choice of the editors of *Wilhelm Dilthey: Selected Works*[1] because this term refers implicitly to the tradition of the "humanities" in the Anglo-Saxon cultural tradition. The same translation has been used in the more recent literature on Husserl's writings though some prefer the originally Neo-Kantian term "cultural sciences" (*Kulturwissenschaften*), which has been used also occasionally by Husserl himself.

The problem with the term *Geisteswissenschaften* is that *Geist* has two connotations in the German tradition. The first tradition has its roots in Hegel's *Phänomenologie des Geistes* and this phenomenology is not a phenomenology of

This chapter was previously published in the following book:
Husserl on the Human Sciences in "Ideen II", in Husserl's Ideen, ed L. Embree, T. Nenon, pp 125–140 (Springer, 2014)

[1] Wilhelm Dilthey. 1989. *Selected Works*, (SW) ed. R.A. Makkreel and F. Rodi. Princeton: Princeton University Press.

T. M. Seebohm (deceased)
Mainz, Germany

© Springer Nature Switzerland AG 2020

T. Nenon (ed.), *Thomas Seebohm on the Foundations of the Sciences*, Contributions to Phenomenology 105,
https://doi.org/10.1007/978-3-030-23661-8_12

the human mind but a historical topology of the appearances of the absolute spirit. The other tradition starts with the translation of John Stuart Mill's *System of Logic.* The usual translation for Mill's term "moral sciences" in German is *Geisteswissenschaften.*[2] Psychology, called by James Mill the analysis of the phenomena of the human mind,[3] is in John Stuart Mill's *System* the foundation of the moral sciences. This is also the case for the relation between psychology and the other *Geisteswissenschaften* in Wilhelm Dilthey's writings. The difference between Mill and Dilthey is that Mill's method in the moral sciences is inductive, experimental, and interested in causal explanations. Dilthey, on the contrary, characterized his psychology as a descriptive and analyzing (*beschreibende und zergliedernde)* science and distinguished in general between the methods of understanding (*Verstehen*) in the *Geisteswissenschaften* and of explanation in the natural sciences.[4]

Boyce Gibson's early translation of *Geisteswissenschaften* in *Ideen I* with "mental sciences" follows the second connotation.[5] The first connotation is, on the contrary, clearly predominant in *Ideen II.*[6] Especially the terminology of the third part and related passages in the appendices indicate that transcendental phenomenology has to be understood as the *absolute Geisteswissenschaft,* the "absolute science of the spirit," and that the human sciences are the spiritual sciences of the *geistige Welt,* the spiritual world. The terminology of *Ideen II* thus prepares Husserl's later use of the terms *Geist* and even *absoluter Geist,* "absolute spirit," as synonyms for the transcendental ego.[7] It is hence advisable to translate *Geisteswissenschaften* in *Ideen II* with "spiritual sciences."

A text ought to be interpreted in its original context following the original intention of the author. The spiritual sciences—explicitly mentioned are only psychology and sociology—are according to *Ideen I* together with the natural sciences as empirical sciences in brackets after the phenomenological reduction.[8] The task of *Ideen II*

[2] J. St. Mill. 1843. *System of Logic: Ratiocinative and Inductive.* London: Harrison and Co; first German translation by J. Schiel, 1849.

[3] James Mill. 1829. *Analysis of the Phenomena of the Human Mind*, ed. J. St. Mill. London: Baldwin and Cradock.

[4] W. Dilthey's method presupposes inner experience and is in this sense positivistic, positivism understood as nineteenth century positivism but not in the sense of the analytic positivism of the twentieth century.

[5] Cf. the translation of *Ideen I,* §60 in: *Ideas. General Introduction to Pure Phenomenology*, trans. W. R. Boyce Gibson (London: Allen and Unwin, 1931). Cf. also *Ideas pertaining to a Pure Phenomenology and to Phenomenological Philosophy*, Book I, General Introduction to Pure Phenomenology, trans. Fred Kersten (The Hague: Martinus Nijhoff, 1982).

[6] Edmund Husserl, *Ideas pertaining to a Pure Phenomenology and to a Phenomenological Philosophy*, Second Book, Studies in the Phenomenology of Constitution, trans. Richard Rojcewicz and André Schuwer (Dordrecht: Kluwer, 1989).

[7] Cf. e.g., *Die Krisis des europäischen Wissenschaften und die transzendentale Phänomenologie* Husserliana VI. Edmund Husserl Gesammelte Werke (Dordrecht: Martinus Nijhoff, 1950), 347, on *absoluter Geist* and *absolute Historizität*; henceforth, Hua VI.

[8] Hua III, 144.

is, according to the outline of the whole project in the introduction of *Ideen I*, the phenomenological analysis of the relation between phenomenology and the natural sciences, psychology and the spiritual sciences, and the a priori sciences. The task of Book III is to develop the idea of a true philosophy as absolute knowledge that can only be realized in pure phenomenology.[9]

The original project of the second volume has been published as *Ideen II* and *Ideen III* in the Husserliana.[10] According to the preface by the editor,[11] the original project of the *Ideen II* had two parts: (A) analyses of the constitution of the material, the animal, and the mental world, and (B) epistemological (*wissenschaftstheoretische*) considerations. (A) Was published following the Landgrebe typescript of 1923–1924 as *Ideen II* in the Husserliana. The "epistemological part" (8) was published in the Husserliana as *Ideen III*[12] with the subtitle "Die Phänomenologie und die Fundamente der Wissenschaften." There are no manuscripts for the third volume of the original project but there are the lectures on first philosophy of 1923–1924 published as *Erste Philosophie I, II* in the Husserliana.[13]

Considering the history of the development from the original project to the selection of the material that has been published as *Ideen II* and *III* both volumes should be read together as parts belonging to the second book of the original project. *Ideen II* can be understood as representing (A) the constitution of material nature, animal nature, and the spiritual world. Reading *Ideen III* as (B) epistemological considerations causes difficulties. To characterize the *Ideen III* as fundaments *(Fundamente)* of the sciences invites this interpretation but the terms that are used in the texts of *Ideen II* and *Ideen III* are "foundation" *(Fundierung)* and "founding" *(fundieren).*[14] Foundations and founding are, however, terms belonging to formal ontology. If so, then investigations on *ontological* foundations of the sciences cannot be understood as *epistemological* reflections in the narrower sense. The published version of *Ideen III* adds reflections on the ontological regions of reality[15] in which first real things and then objects of the sciences are given[16] and on the relation between phenomenology and ontology in general.[17] Material nature and natural sciences are mentioned.[18] Much is said about psychology but virtually nothing about the other

[9] Hua III, 7–8

[10] Hua IV and V. The original textual basis of the editions of *Ideen II* and of *Ideen III* is the collection of manuscripts of Edith Stein 1913f. in the typescript of Landgrebe 1923–1924. Landgrebe's version of *Ideen II* together with later additions and notes of Husserl is the material that has been used in the edition of *Ideen II* in the Husserliana. Cf. Hua IV, Introduction of the Editor, XVIII.

[11] Hua IV, XIV.

[12] Hua IV, XIV f.

[13] Hua VII and VIII.

[14] Cf. e.g., Hua V, 17. The real things are the foundations of the objects of the sciences.

[15] Cf. *Ideen I*, Hua III, §72, §153.

[16] Hua V, Chapter I.

[17] Hua V, Chapter 3.

[18] Hua V. Of significance is the remark on p. 9 that biology dealing with animals and plants and organic life is a part of material nature. Nothing more is said about the life sciences in *Ideen III.*

spiritual sciences. What *Ideen III* say about the foundations of psychology as rational, i.e., eidetic psychology as a spiritual science can also be understood as first step towards the investigations in the lectures on *Phänomenologische Psychologie.*[19] *Ideen III* is, however, not of significance for the analysis of the constitution of the spiritual world as the foundation of the spiritual sciences in Parts III and the appendices of *Ideen II*.[20]

The remarks about Dilthey in Part III and the appendices are of crucial significance for the understanding of *Ideen II* in the context of the discussion of the *Geisteswissenschaften* in Germany in the beginning of the twentieth century. The remarks have a pre-history. Following Stumpf, Husserl characterized his method in the first edition of the *Logische Untersuchungen* as *descriptive-psychologisch*, i.e., as the method of a descriptive psychology. Dilthey called his psychology a descriptive and analyzing psychology. He praised Husserl's *Logische Untersuchungen* as a partial realization of his own projects[21] and used occasionally the term "phenomenological" to characterize his method.[22] He was therefore deeply offended by Husserl's critique of historical relativism in the third section of "Philosophie als strenge Wissenschaft" because he assumed that he was the target of this criticism and defended himself in a letter to Husserl 1911. Husserl's answer emphasized that Dilthey was by no means the target but also that he and Dilthey would be able to agree in all questions of the interpretation of the human mind and the analysis of the problems of understanding of cultural objects.[23] Husserl's repeated references to Dilthey can be understood as an extension of his answer and an indication of Dilthey's significance for the development of Husserl's reflections on the spiritual sciences in general.[24]

---

[19] Hua IX.

[20] There are early manuscripts about the cognitive attitude of the spiritual sciences (*geisteswissenschaftliche Einstellung*) written in connection with the original project between 1913 and 1917 that have been dropped from the draft of the main text in Landgrebe version of 1923/1924 and added as appendices by Landgrebe together with further remarks from Husserl himself. Cf. Hua IV, esp. Appendices IV, V, X, XL XII part II, XIV and the remarks to the appendices of the editor, 417f.

[21] Wilhelm Dilthey, *Gesammelte Schriften*, Vols 1–12 (Teubner, 1914ff.); Vol. VI, *Die geistige Welt. Zweite Häfte* (1924), 13f., 39f., 322.

[22] GS VII, *DerAufbau der geschichtlichen Welt in den Geisteswissenschaften* (1927), 296, "phenomenology of knowledge."

[23] Central passages of the letters can be found in *Phänomenologische Psychologie,* Hua IX, Preface of the editor, XVII–XXI. English trans., Edmund Husserl, *Phenomenological Psychology*, trans. John Scanlon (The Hague: Martinus Nijhoff, 1977).

[24] It is not possible to discuss the remarks about Dilthey point by point in this essay. The main references to Dilthey can be found in Hua IV, §48: Introduction to section III; appendix XII, §11,365 ff. and XIII, 393; and Hua IX, *Phänomenologische Psychologie*§1, 5–6, 10; §2, esp. 13–14, and Appendices II, 357 and III, 361. On Husserl and Dilthey, cf. *Dilthey und die Philosophie der Gegenwart,* ed. E.W. Orth (Freiburg: Karl Alber, 1985), the essays of Ströker, Seebohm, Pfafferott, Makkreel, and Carr and *Dilthey and Phenomenology,* eds. R. A. Makkreel and J. Scanlon, Current Continental Research 006 (Washington, D.C.: CARP & University Press of America, 1987), Part I: Dilthey, Husserl, and the Foundations of Science.

Husserl praised Dilthey for his radical critique of the claim of the naturalistic, i.e., psycho-physical, physiological, and experimental psychology of the nineteenth century and his ability to explain the facts of mental life in his *Ideen zu einer beschreibenden und zergliedernden Psychologie* of 1894 as a work of a genius. The theory of understanding (*Verstehen*) as the basic category of Dilthey's psychology is not only able to analyze the laws of psychic life but together with them also the laws of interpersonal mental life, of culture and regions of mental culture, i.e., religion, the law, literature, and fine art. Understanding psychology is, hence, the basic science and foundation for all other mental sciences. What is missing in Dilthey is, however, precision in elementary analyses of experience and a distinction between the constitution of the psychic reality in the constitution of the animated living body and the constitution of the mental world.

The main shortcoming is Dilthey's assumption that the empirical generalizations of a morphological typology based on inner experience and the inner reliving of foreign mental life is able to discover the universal laws of psychology and cultural spiritual life. Only the method of eidetic intuition is able to discover the essential structures of psychical and spiritual life and its cultural manifestations. Dilthey is not able to give a precise analysis of interpersonal spiritual life and, therefore, also not able to give a satisfying account of the specific objectivity of the manifestations, the works of spiritual life.

What is missing in Husserl's critical appraisal are references to Dilthey's epistemological investigations of the methods of the historical spiritual sciences and their background in the methodological reflections in the nineteenth century.[25] More will be said about this problem in some critical remarks in the end of this chapter.[26]

Part III of *Ideen II* on the constitution and ontological priority of the spiritual world presupposes Part II, the constitution of the animal world and Part II presupposes section I, the constitution of the material world. The considerations of

[25]The main source for Dilthey's epistemological reflections on the spiritual sciences is *Der Aufbau der geschichtlichen Welt in den Geisteswissenschaften* first published 1910 and then in GS VII; English trans., *The Formation of the Historical World in the Human sciences,* SW/11, (Princeton, NJ: Princeton University Press, 2002). The question is whether Husserl ever read the *Aufbau.* There is no trace of such a reading in the material in *Ideen II* or *Phänomenologische Psychologie.* Husserl read and mentioned (Hua IX, 6) Dilthey's *Ideen zu einer beschreibenden und zergliedernden Psychologie* of 1884 (now in GS V 1924) and his remarks to Dilthey indicate that he also knew the *Einleitung in die Geisteswissenschaften* of 1883 (now GS I 1922 and SW I, *Introduction to the Human Sciences).*

[26]On the methodology and epistemology of the human sciences from Schleiermacher and Boeckh to Dilthey and Droysen, see Makkreel, R.A. 1975. *Dilthey: Philosopher of the Human Studies.* Princeton: Princeton University Press, esp. Part Three; Seebohm, T.M. 1984. Boeckh and Dilthey: The Development of Methodical Hermeneutics. *Man and World* 17: 325–346, and Thomas M. Seebohm, *Hermeneutics. Method and Methodology.* Contributions to Phenomenology 50 (Dordrecht: Kluwer Academic Press, 2004), Chapter 3.

Ideen II on the spiritual sciences have, therefore, to begin with a short summary of section I and a survey of the basic theses of section II.[27]

The real things of real nature are given in the natural attitude as material real things, animal things, and as things that are of value for practical life, valuable things.[28] The objects of the natural sciences are theoretical objects. The givenness of theoretical objects has its foundation the givenness of real things in aesthetic synthesis (§18 a–c). Already the constitution of real things given in the natural attitude and, therefore, also the constitution of theoretical objects presuppose the subject not as a solipsistic subject but the subject as a member of an intersubjective community (§18 e, f, and h). The givenness of the theoretical objects of the natural sciences presupposes an abstractive reduction that brackets all aspects of real things as things that are relevant for practical life (25f.). What is left are objects as facts (*Sachen*). The theoretical objects of physics, the pure material objects require in addition an abstraction from all secondary qualities of real things and theoretical objects. What is left are the primary qualities, i.e., measurable and in this sense mathematically quantifiable qualities.[29] This second abstraction excludes in addition all essential properties of animal nature.

Part II begins in the introduction with a description of the givenness of the souls of humans and animals in the natural attitude. What "animal nature" means is, hence, "animated nature." The soul is originally given together with and in the material living body but it has its own unity as an endless stream of lived experiences without beginning and end. The I is a unity within the unity of a real soul as an I with a certain character and certain abilities. It is this I that *has* a body and *has* a soul (§21, 93/94). This structure, including its foundation in the givenness of the material living body, is the foundation for the givenness of others as other living bodies, empathy and the constitution of intersubjectivity (95).

The analysis of the constitution of this structure presupposes in a first step the "abstraction" from the givenness of the living body in the natural attitude that is implied in the phenomenological reduction. What remains is the pure spiritual ego as cogito and its cogitations. The detailed analyses of the constitution of animated nature therefore starts in Chapter I with an extension of the analyses of *Ideen I* that can also be read as a first sketch of the *Analysen zur passiven Synthesis* and their

---

[27] Two comprehensive essays on *Ideen II* have to be mentioned: Alfred, Schutz. 1966. Edmund Husserl's *Ideas*, Volume II. In *Collected Papers III*, *Studies in Phenomenological Philosophy,* ed. I. Schutz. The Hague: Martinus Nijhoff and Paul, Ricoeur. 1967. Husserl's *Ideas II*: Analyses and Problems. *Husserl: An Analysis of His Phenomenology*, trans. E.G. Ballard and L. Embree. Evanston: Northwestern University Press.

[28] (8/9; 25) The references to paragraphs, numbers of appendices, and/or page numbers of *Ideen II,* Hua IV will be given in parentheses on the following pages.

[29] §18 d and g. What is said there can be read as a first sketch of what will be said later in the *Crisis* about the abstraction that is a necessary presupposition for the theoretical attitude of the natural sciences in general, cf. e.g., Hua VI, §2, esp. 3 and 4, §66, 230 and §§8–10 about the mathematical abstraction presupposed in the theoretical attitude of physics.

application in the "V. Cartesianische Meditation."[30] The task of the following investigations is the analysis of the foundations of the understanding of human beings as persons with their personal environment as self-subsisting individuals in the spiritual world of customs, laws, religion etc.(§34, esp. 141).

The first step is the analysis of the constitution of the reality of the living body as an organ of the solipsistic soul. The living body is given in bodily localized sensitivity, *Empfindnis*, and as an organ of will in the kinaesthetic movements of the living body (§§36–38). The living material body is furthermore a center of orientation, a "Here" opposed other material things in a manifold "There." This localization of the givenness of the living body is not the givenness of the living body in objective real space. It is a solipsistic system of localization as one material thing among others given in "causal connections," the sensitive experience of the spontaneity of bodily actions and the passivity in the experience of feeling conditioned by the presence of real material bodies in the There outside. The constitution of the solipsistic living body is, hence, given in the inner perspective of the sensitivity of the living body in the flow of psychic life. It is also given simultaneously in the outer perspective as a center that is a correlate of a world in the There outside the living body that conditions the inner experience and can be conditioned by the activities of the living body as an organ of the will (§§39–42).

What is originally present in the sensory experience of the inseparable unity of solipsistic body and soul are (a) the material things given in the There outside and the partial givenness of the own solipsistic living body from the outside as a material thing and (b) the individual and unique flow of inner experience in which the solipsistic subject is given for itself. What is not yet given in solipsistic experience are the material thing given as an independent real thing in real space and other living bodies as animated bodies in the natural attitude. Animals, i.e., animated material bodies of humans and non-human animals, are given beyond solipsistic experience as bodies in original presence (*Urpräsenz*) as an appresented psychic inner life. Empathy (*Einfühlung*) is beyond simple appresentation an intentional attitude of an ongoing steady appresentation—but not presentation—of their foreign inner live given with the experience of the presence of their living bodies (§§43–45).

Empathy founded in appresentation is constitutive for the givenness of intersubjectivity and with it for the constitution of the human ego *(lch-Mensch)* given in the natural attitude. Intersubjectivity is, as already mentioned in Part I, the foundation for the experience of real material things and real material nature in the natural attitude. The aesthesiological constitution of real things in the natural attitude is the foundation for the theoretical objects of the natural sciences that can be given after the reductive abstraction that brackets all the characters of real things given in intersubjective practical life. The attitude of the natural sciences presupposes, hence, a

[30] Cf. Hua IV, pp. 99/I 00, 104 and Hua III p. 136/7, the *Analysen zur passiven Synthesis,* Hua IX, Parts III and IV, and the *Cartesianische Meditationen,* Hua I, §§44–46. Edmund, Husserl. 2001. *Analyses Concerning Passive and Active Synthesis*, trans. Anthony J. Steinbock. Dordrecht: Kluwer Academic Publishers..

field of an intersubjective experience in which real things, material as well as animated real things, are given (§§46, 47, esp. 171–72).

The field of intersubjective experience in which nature and the subject itself is given in the spiritual world is the realm of objects in the theoretical attitude of the spiritual sciences. The reference to Dilthey in the introduction to Part III (§48, 172–73) indicates that the following analysis of the personalistic attitude can be understood as the phenomenological explication of Dilthey's intentions. Dilthey's understanding, descriptive and analyzing psychology based on inner experience is for him the foundation for all other disciplines of the spiritual sciences. *Ideen II* replaces Dilthey's basic category of understanding[31] and its foundation with the descriptive analysis of the difference between introjection and empathy. The foundations of empathy in the constitution of the solipsistic animate body (§§43–45) are the foundations of the personalistic attitude and the distinction between the naturalistic and the personalistic attitude. The personalistic attitude is the foundation for the analysis of the spiritual attitude, the spiritual world, and the spiritual sciences.

The analysis of the constitution of the spiritual world starts with a summary of the analysis of the constitution of animated living bodies in the naturalistic attitude. The living body is in this attitude given as physical real object. The soul is given as a layer of real events that can be localized on a living body, i.e., as the aesthesiological layer of sensations and sensitive experience on living bodies. The inner temporality of lived experience is understood as a phase of the objective space-time. This givenness of the other soul can be characterized as "introjection," but this introjection is not directed to something behind the physical givenness of the living body (§49 a–c). The naturalistic attitude is, however, only a possible attitude because it presupposes an abstractive reduction and has its foundation in the natural world given in the natural attitude (§49 d and e). The other possible attitude is the personalistic attitude and with it the attitude of the spiritual sciences.

The subject is given in the personalistic attitude as a person in the center of its personal environment. The world is given as the world of the consciousness of the individual personal I. The I as person, the egoic person, is given in this world in its relations to the real things as a representing, feeling, evaluating, striving, and acting person. The world is given as experienced world in sensory appearances. The real things in the world are given in the appearances as valuable things, i.e., as pleasant, painful, beautiful and ugly things in immediate experience. The person is related to the real things in practical acts. Practical needs and desires are the motivations for practical acts. What is desired are useful objects that can be used in practical acts to produce valuable artifacts *(Erzeugnisse)*. The relations between things and persons are in the personal attitude not given as causal relations. They are given as motivations, enticements of more or less powerful tendencies for evaluations, practical activities, and finally theoretical judgments about the pleasant or painful, beautiful or ugly, and useful character of objects. The relations of the person to real things are, seen from the viewpoint of pure phenomenology, intentional acts (§50).

[31] See, however, the critical remark in fn. 40 about a misinterpretation of Dilthey's distinction in Appendix XIII, 393.

The person considers itself, its relations to real things, and its motivations in the personal attitude as a member of a personal community. Other persons are given in a personal community (*Personenverband*). They are not given as introjected in a material living body: They are given immediately in empathy as persons together with the real things in the common environment of a personal community. Persons are given in the personal community as independent subjects with their own rights in a system of moral and legal relations in which each person is a person only because it is recognized as a person by other persons as a subject-object. The egoical person is motivated in the personal community to reach agreements with other persons and is, therefore, influenced by other persons. This system of reciprocal influences, agreements and disagreements, is constitutive for the givenness of the common environment including not only a real environment of physical things and animals but also an environment of recognized ideal entities.

The egoic person has to pursue its own goals in the environment of the personal community. The social community is in turn open for actual and potential communication with external environments of other personal communities, other spiritual contexts in its environment. This openness is constitutive for the idea of the world as spiritual world, the totality of different types of real and ideal objects (195 f).

Real things are given in the different perspectives of different personal communities in the spiritual world. The spiritual world, though not apprehended as objective physical nature in the naturalistic attitude, has as its medium a common intersubjective, i.e., objective space-time in which the different personal communities and finally the individual persons are located in the spiritual world.

Real things and real nature can be determined as the true objects and true nature in itself[32] behind their appearances for different personal subjects and personal communities in a theoretical attitude. They are constituted in this attitude as theoretical objects that cannot be given in sensory perspective variations. This theoretical attitude is the naturalistic attitude of the natural sciences. The investigation ends at this point in a vicious circle (210). The starting point was the apprehension of humans in their physically given living bodies in the naturalistic attitude and ended with the analysis of the givenness of the soul. The analysis of the flow of lived experience ended with the turn to the personalistic attitude and the constitution of the spiritual world. The outcome of this analysis was, however, in the end the recognition that the world given for universal intersubjectivity as the spiritual world can also be given as the correlate of universal intersubjectivity as the world as correlate of the naturalistic attitude.

A solution for the problem of the circle requires (§49) a detailed recapitulation of the analysis of the structures of the personal I given in the *inspection sui* in self-awareness and self-consciousness. Self-consciousness is the reflective awareness of the not reflected consciousness of that what is pre-given as the Not-I for the personal self (211–12; cf. 248). The I is in its self-awareness not given to itself as an I *in* the living body. It is aware of itself as *having* a body and this body is, on the one hand,

---

[32] They are intersubjectively given not for a relative community but as intersubjective in principle; cf. *Ideen I*, Hua III, 113.

an organ of the bodily activities of the I, reacting to pre-given real things in acts of attention, judging, evaluating, and willing, and as the decision to realize change in the pre-given environment of the I. The I as having the body is, on the other hand, also given as the passive I, the I of receptivity, of being attracted or repulsed by stimuli *(Reize)* from presence or absence of real things, the feelings accompanying such stimuli, and finally by the self-awareness of the bodily activities of the I.

The active and passive relations between I and Not-I are real relations and real relations must be distinguished from intentional relations between the personal I and things as intentional objects. Intentional relations can be given together with the real relations between the I and real things but they and their intentional objects can also be given after or even without real relations. Something can be my intentional object as a noema without being really present in a real relation.

Relations between real things in real nature, of objects (*Sachen*) given in the naturalistic attitude, including the mathematically interpreted theoretical objects of physics, are *causal* relations (216, cf. 230–31). Considering the relations between the personal subject and its environment including real things but also ideal objects and noematic objects in general *because* referring to these relations is understood as the because of *motivations.* Stimuli of sensory impressions motivate the subjective-objective effects in the passive awareness of objects. They motivate furthermore feelings of pleasure and pain of the I and active goal-directed reactions of the body as an organ of the I. What can be distinguished are the motivations following the norms of reason, i.e., of logical thinking and evaluations and secondly motivations determined by associations belonging to the sphere of passivity and habits. Lived experience is thus a web of motivations that are partially active and conscious and partially passive and often not immediately accessible for explicit reflections. Even the determination of every temporal phase by the preceding phase in the structure of immanent time consciousness can be characterized as a type of passive motivation (225, cf. 227).

The experience of others is in the naturalistic attitude a theoretical *introjection* of the soul as a dependent appendix of a material living body. The experience of other persons in the personal attitude is the immediate experience of the other person as the subject of its motivations in its environment of real things and other persons (228–29). The living body of the other person is understood as an organ of the bodily actions of the other person, as expressions of its motivations and actions (234–35). Persons in communities of persons are understood as subjects of motivated actions that are determined in reciprocal motivations and not as causal relations. The "because" of *motivations* of others is like my own motivations immediately understood and this *understanding* is not in need of further causal explanations, much less causal explanations defined in terms of mathematical theories (230–31; 235–36).

The spiritual world includes other persons, the own personal community and external personal communities, but it includes also (239, 243–44, cf. 182, 195 f) cultural objects: artifacts produced for daily use (*Gebrauchsgegenstände*),written texts (*Schriftwerke*), works of art, musical works, scientific and legal literature, symbols, etc. (239, 243–44, cf. 182, 195f.). The "body" of such works of the spirit is physical but it is a body of a spiritual meaning. The physically given signs, e.g., letters, sentences are animated (*haben Beseelung*). They intend an ideal object.[33] The spiritual meaning of the signs and their sensory sign matter are not separable (238). They are partially real but they are signs only because they mean ideal objects. They are objective spirit as manifestations of the spirit (239; 243). The mental world is, hence, (1) the world in which the person finds itself in introspection *and* in the representations of the other person(s) in empathy. It is (2) then world of communities of higher order, the spiritual world, including other communities that have on different levels different common spirits, the spirit of a nation, of a culture, a religion etc. (3). This universe is accessible in the manifestations of spirit.

The task of chapter 3 of Part III on the ontological priority of the spiritual world to the naturalistic world (281f.) is to resolve the circle mentioned above (210). The first step is to show that the circle is implied in the assumption of the parallelism of the spiritual and the naturalistic world. The second step is the refutation of the parallelism.

The spirit presupposes the soul given in the flow of lived experience and the soul is dependent on the living body for the personalistic attitude, i.e., the attitude of the spiritual sciences. The activities of the spirit are, hence, dependent on the living body as physical body and the works of the spirit are likewise as objects given as physical objects in the world of the naturalistic attitude. They are, however, works of the spirit only because they have a spiritual meaning. The spirit acts on nature but these actions *(Wirkungen)* have not the character of causal relations between natural objects. They are of significance only for personal communities and their environment.

The transition from the attitude of the spiritual sciences to the naturalistic attitude presupposes an abstraction. The abstraction excludes the spiritual attitude and the spiritual world. The animated body and the soul are not given for a person who has a soul and a body. The body is given in the naturalistic attitude as a physical body and the soul is given as introjected in the body.

Seen from the viewpoint of the spiritual attitude, the natural sciences appear as a work of the spirit. However, seen from the viewpoint of the theoretical I, the I in the theoretical attitude,[34] the transition to the naturalistic attitude and the implied abstraction is an act of the theoretical I, but the attitude of the spiritual sciences implies *also* an abstraction, the abstraction from the correlate of the naturalistic attitude the physical world. There are, hence, two parallel worlds, the world of the spiritual sciences and the world of the natural sciences for the theoretical I (289).

The presupposition of this parallelism implies (a) the assumption that every event in the world of nature has its correlate in the spiritual world and vice versa and

[33] It is, according to *Ideen II*, 243, not difficult to give a detailed analysis. However, such an analysis would be the epistemological analysis of the "methodology and encyclopaedia," e.g., of the science of classical antiquity according to August Böckh and this is by no means a simple enterprise, cf. below fn. 49.

[34] The theoretical I ought not to be understood as the transcendental ego. The theoretical I is the I of the theoretical attitude of the natural as well as the spiritual sciences.

happen at the same time and (b) the additional assumption[35] that both events are determined by reciprocal causal connections. The usual assumption is that the "place" in which that happens is the central organ, the brain (290–294). A phenomenological analysis of temporality is able to reject assumption (a). Events in the central organ follow each other in the structure of objective time. Events in the lived experience happen, in the structures of inner time consciousness. The hyletic contents emerge in the actual now in its protentional horizon and flow down in the continuum of retentions. Given this distinction it is also impossible to understand the reciprocal conditioning, the acting *(das Wirken)* of the soul and the spirit on the living body and vice versa as a causal relation. Causal laws are the conditions of the course of events in the naturalistic world. What conditions events in the soul and the spiritual world are motivations and motivations can be immediately understood. Causal relations are discovered as causal laws in the naturalistic attitude with the aid of experimental research (295–96).

The spirit depends partially on nature but this nature is the nature of the natural attitude and not the physical nature of the naturalistic attitude. If we strike out the spirit, no nature is left. But the spirit as isolated individual spirit and its motivations without a personal community is left if we strike out nature (207). Individuality in the spiritual sense is absolute individuality and has its motivation in itself. It is the *heacceitas* in which concrete essence and existence cannot be separated (300). The individuality of real things in the natural attitude is a relative individuality and determined by external causality. It is an open individuality that emerges and can vanish in the course of events. Objective nature is a correlate of the naturalistic attitude and not a pre-given transfinite whole in the actual Now of lived experience. It is only given in the progress of the research of the natural sciences that is open for revisions and corrections (299).

The appendices offer in most cases other versions of the analyses in the main text. There are, however, also (a) additions considering a possible transition from the spiritual attitude to the transcendental phenomenological attitude and (b) one addition in which the spiritual and historical world is characterized as lifeworld. There are finally (c) additions that are relevant for the methods and the system of the spiritual sciences.

(a) Psychology is the basic discipline of the spiritual sciences first as empirical descriptive psychology and then as eidetic psychology. Reflections on the presuppositions of psychology prepared the path to the idea of a pure phenomenology, the eidetic and then the transcendental reduction in the *Logische Untersuchungen.* This first approach has epistemological and philosophical shortcomings. Reflections on the presuppositions of the empirical spiritual sciences in general and the structures of the spiritual world in eidetic intuition include all regional ontologies and all a priori disciplines. They offer a new and comprehensive path to pure phenomenology (IV, 313–14).

[35] It is an assumption because there are other assumptions, e.g., the metaphysical assumptions of the occasionalists following Geulincx and Malebranche rejecting a possible Cartesian *influxus physicus* in the brain and assuming "occasional" interventions of God.

As understanding sciences, the spiritual sciences presuppose empathy. They are empirical positive sciences of persons, social communities, and the formation and genetic transformation of persons and social communities. Real things and real nature are reduced to constitutive unities given in intersubjective regularities (XII, II. 348, 363–64, 369–70).

The naturalistic attitude presupposes an abstractive reduction. No abstractive reduction is required for the transition to the spiritual attitude. The world of the spirit world is the world of the natural attitude, the world of the original experience of persons and their social and natural environment. Transcendental phenomenology is the absolute spiritual science. Only one step is required for the transition from the spiritual attitude to transcendental phenomenology, the phenomenological reduction (XII, 354, cf. 366, 369).

A person has a history and this history implies in its horizon the historical world of past and present cultures. Historiography is the description of true history as a consistent experience from the viewpoint of the actual historical presence of the historian. History is (a) originally accessible in empirical descriptions of the environment of persons and social communities. It is (b) explicable in a universal empirical morphology on the level of objective science and it is finally (c) explicable in a priori descriptions of the essential structures (*Wesensformen*) of persons, their environment, and their historical horizon (XII, 11, 353–55, 371–72).

The history of the natural sciences is as history a special discipline of the spiritual sciences but that does *not* imply that real nature is an object of the spiritual sciences. Nature is an object of the natural sciences and persons as living bodies as well as all other spiritual objects are accessible for the natural sciences as physical objects. There is thus not only an opposition but a remarkable parallelism between the natural and the spiritual sciences.[36]

(b) Of significance for the further development of phenomenology is that the term "natural world" in appendix is replaced by the term "lifeworld" in appendix XIII and what is said about history in appendix XII is said about the lifeworld in appendix XIII.[37] The spiritual world was characterized in XII as the world of the natural attitude. Lifeworld is, hence, a new term for the world given in the natural attitude.

(c) Setting aside the passages about psychology and history and their methods and a remark about linguistics, *Sprachforschung* (XII, 367), no other disciplines of the human sciences are mentioned in *Ideen I*. An outline of a system of the objects of the spiritual sciences can be found only in Appendix V. Mentioned are (V, 315 f) (1) the general structures of the natural environment of humans and human communities, (2) Humans and animals in the environment,(3) Commodities (*Güter*), works of art like artifacts, tools etc., and (4) Customs and

[36] XIV pp. 390 ff, esp. 392–93. What is said in appendix XIV, probably from 1917, is at least *prima facie* not compatible with the refutation of this parallelism in chapter 3 of Part III. Cf. the notes to the text of the editor, pp. 423–24.

[37] 373–75. Both texts were written in 1917. Cf. editorial remarks, pp. 418, 423.

moral principles, laws, language etc. i.e., in general the systems of norms and conventions that determine social personal communities. Communities that are communities without a will and goal directed actions are called social personalities, e.g., linguistic or ethnic communities.

Objects given in spiritual life have a genesis. There are (1)[38] biographies interested in the genetic developments of individual persons and (2) different types of histories, (2.a) the histories of concrete social communities, towns, nations etc.; (2.b) histories of the arts and the sciences and other irreal (ideal) objects; (2.c.) histories of social change and the development of mankind in general and including the pre-historical origin of humanity and its animal-like ancestors.

12.2 Concluding Remarks

Ontological foundations of the sciences are regional ontologies. The ontological apriori of empirical sciences can be discovered in morphological reflections on the empirical material and beyond that in eidetic intuition. The transcendental phenomenological justification of this a priori requires the analysis of the constitution of eidetic intuition and of the regional ontologies.[39]

The task of *Ideen II* was the analysis of the constitution of the ontological foundations of the sciences. The task of *Ideen III* was the analysis of these foundations.[40] What *Ideen II* offers is the analysis of the constitution of the different foundations of the natural and the spiritual sciences. The phenomenological analysis was able to justify Dilthey's intuitions and to eliminate its shortcomings.[41] What can be found in *Ideen III* is an explication of the ontological foundations of psychology.[42] An analysis of the constitution of personal communities and the social lifeworld can be found in *Ideen II.* Nothing is said in about the foundations of the other social sciences in the *Ideen III.* Apart from the analysis of the constitution of intersubjectivity, especially as presented in the *Cartesianische Meditationen*,[43] not very much

[38] The enumeration in the following list is added.

[39] *Ideen III,* Hua V, Chapters 1, 2 §§5–7, Chapters 3 and 4, cf. Appendix I, §6.

[40] Some remarks about the ontological foundations of physics can be found in Chapter 2, §11.

[41] Cf. above fn. 30, and about the references to Dilthey, §48 of *Ideen II* in the main text. The critique of Dilthey's distinction in Appendix XIV, 393 is incompatible with what is said in §48 and with Dilthey's distinction between the natural sciences as sciences of causal explanations and the spiritual sciences of understanding. XIV claims that Dilthey characterized the spiritual sciences as descriptive sciences and also points out that the natural sciences presuppose descriptions. Dilthey characterizes his psychology as descriptive, but this description presupposes, like all other spiritual sciences, understanding, i.e., precisely that what is analyzed as empathy in *Ideen II.*

[42] What is said in *Ideen III* is a first sketch of the detailed analyses in the *Phänomenologische Psychologie*, Hua IX and the *Analysen zur passiven Synthesis,* Hua XI.

[43] Hua I, §§42–47.

beyond that can be found in Husserl's later works, but it can be assumed that Husserl thought that the work of Alfred Schutz would fill this gap.[44]

The eidetic structures of history or historiography[45] presuppose genetic constitution.[46] The genetic aspect of history together with the analysis of the animal nature and its significance for the spiritual world admits the recognition of the animal origin of the historical development of mankind in the context of the spiritual sciences.[47] Nothing beyond that is said about the constitution and the foundations, set aside the epistemology, of the historical spiritual sciences. There is no evidence in *Ideen II* that Husserl ever read Dilthey's *Der Aufbau der geschichtlichen Welt in den Geisteswissenschaften.*[48] Dilthey's distinctions between the historical and the systematic or social spiritual sciences, between elementary and higher understanding, between pre-methodical and methodical understanding, and between immediate and fixed life expressions are not mentioned. There is also no evidence that Husserl was familiar with the background of Dilthey's epistemology of the spiritual sciences in the development of the reflections on the methodology of the philological historical method, hermeneutics, and histories in the literature about hermeneutics of the nineteenth century.[49] There is no word about Schleiermacher's first and second canon of hermeneutics or about Boeckh's reflections on the methodology of philology and his distinction of grammatical, individual, historical, and generic interpretation and critique.[50]

The task of *Ideen II* is the analysis of the constitution of the ontological foundations of the empirical sciences. This analysis is the presupposition for but it does not yet include a phenomenological epistemology of the empirical sciences. It is, hence,

---

[44] Alfred Schutz was dissatisfied with the analysis of the constitution of social groups, cf. pp. 38–39, fn. 26. In his review of Husserl's *Méditations Cartésiennes* (1931) in the *Deutsche Literaturzeitung*, December 18, 1932, Schutz writes "To Husserl's list I would like to a social science which, while limited to the social sphere, is of an eidetic character. The task < of such a social science > would be the intentional analysis of those manifold forms of higher-level social acts and social formations that are founded on the—already executed—constitution of the alter ego. This can be achieved in static and genetic analyses, and such an interpretation would accordingly have to demonstrate the aprioristic structures of the social sciences." Trans. Helmut Wagner in Alfred Schutz, *Collected Papers*, Vol. IV, ed. Helmut Wagner, George Psathas, and Fred Kersten (Dordrecht: Kluwer Academic Publishers, 1996), 164.

[45] Both terms are used as synonyms. The methodologies of hermeneutics and histories of the nineteenth century distinguished historical research and historiography as representations of the results of historical research. See *Hermeneutics Method and Methodology* l.c., fn. 25, §10 on Droysen.

[46] Husserl distinguished later between genesis as subjective genesis and generation as the genesis of intersubjective communities in manuscripts 1929–1936 published as *Zur Phänomenologie der Intersubjekivität III,* Hua XV.

[47] Appendix V, 316.

[48] Cf. above, fn. 24 and 25.

[49] Cf. §§8, 9, n. 44.

[50] Philology, and its methodology, hermeneutics, the basic discipline of the philological historical method and histories is not mentioned in *Ideen II*. What is mentioned is linguistics *(Sprachforschung)*, but this covers only the lowest level of Boeckh's methodology, grammatical interpretation and critique.

not a shortcoming of *Ideen II* that nothing is said about the problems belonging to the epistemology of the spiritual sciences and especially the historical spiritual sciences.[51] The theoretical cognitive attitude of empirical sciences in general brackets the practical interests of the lived experience in the present lifeworld.

It has, however, roots in basic needs of the lifeworld.[52] This root is in case of the spiritual sciences the practical need to avoid misunderstanding and not-understanding in the process of understanding in communications. The first step of a phenomenological analysis of the epistemological problems of the spiritual sciences is the descriptive analysis of the difference between the temporal structures of the understanding of bodily expressions of other persons in the present and the physically fixed "works of the spirit" of authors in the past. This step determines the epistemological distinction of the historical and the social spiritual sciences. The historical sciences are restricted to the interpretation of the physically fixed "works of the spirit" of authors of the past. Misunderstandings and failures of understanding in present communications can be eliminated in dialogues.[53] The interpretation of fixed life expressions of authors in the past need methodologically guided interpretations. The first methodological canon for such interpretations requires that a text has to be interpreted in its own and not the context of the interpreter, i.e., the canon implies a methodological abstraction.[54] Historical research presupposes such interpretations in the reconstruction of "what really happened" in a past phase of a foreign historical development.[55]

Thomas M. Seebohm was professor emeritus at the Johannes-Gutenberg Universität Mainz. For a more detailed description of his career and his work, please consult the "Introduction" to this volume.

[51] A phenomenological epistemology of the empirical sciences requires more than the phenomenological givenness of the ideal objects of mathematics. The distinction of adequate and inadequate evidence is not sufficient because there is no flawless adequacy of evidence, e.g. of presently recognized "laws of nature." Knowledge in the empirical sciences remains fallible. This must be added to the investigation of Zhongwei, L.I. 2010. Towards a Husserlian Conception of Epistemology. In *Advancing Phenomenology. Essays in Honor of Lester Embree,* ed. T. Nenon and P. Blosser. Dordrecht: Springer, sect. 4 and 5, pp. 124f.

[52] Practical life is vitally interested in successful predictions of events following certain actions. This is the pre-scientific root of the theoretical interest of the natural sciences in causal connections and of the development of a scientific technology.

[53] Cf. §§33–35, l.c.: 44.

[54] Cf. §§27 and 36, l.c. 44.

[55] The first canon implies a methodological abstraction. The philologist and the historian has to "bracket" the prejudices of the own historical context. The "uninterested observation" of research in the social sciences requires also an abstraction. It is, hence, a mistake to assume that the methodology of the human sciences as sciences does not imply methodological abstractions.

Chapter 13
Fichte's and Husserl's Critique of Kant's Transcendental Deduction

Thomas M. Seebohm

Abstract The specific topic of this chapter is the difference between the attempt in speculative and dialectical thinking on the one hand, and transcendental phenomenology on the other, to solve the enigmas presented by Kant's transcendental deduction. The thesis is that they are diametrically opposed. The main concern is systematic and not philological-historical. That means, among other things, that the well-known fact that Husserl has a certain preference for the deduction in edition A and that Fichte refers mostly to edition B will not be corroborated in an interpreting of all the passages in both in which they refer to the deduction. What is at stake is a general systematic and theoretical explanation of this and other facts due to the fundamental difference in the foundations of their projects and their understandings of nature of transcendental philosophy.

The literature concerning the relation between Fichte and Husserl, though far from exhausting all possible historical and systematical aspects, has pointed out parallels and differences.[1] The specific topic of this essay is the difference between the attempt in speculative and dialectical thinking on the one hand, and transcendental phenomenology on the other, to solve the enigmas presented by Kant's transcendental deduction. The thesis is, that they are diametrically opposed. The main concern is systematic and not philological-historical. That means, among other things, that

[1]Mohanty, J.N. 1952. Fichte's 'Science of Knowledge' and Husserl's Phenomenology. *Philosophical Quarterly* (Calcutta) 25:113–125; Hyppolite, J. 1959. Die Fichtesche Wissenschaftslehre und der Entwurf Husserls, *Husserl et Ia pensee moderne,* 182–189. Den Haag; Schneider, P.K. 1965. *Die wissenschaftsbegründende Funktion der Transzendentalphilosophie.* Freiburg/München, Chapters 6 and 7; Rockmore, T. 1979. Fichte, Husserl and Philosophical Science. *International Philosophical Quarterly* 19:15–27; Tietjen, H. 1980. *Fichte und Husserl: Letztbegründung, Subjektivität und praktische Vernunft im transzendentalen ldealismus.* Frankfurt a.M; cf. the review of this book by Schuhmann, K. 1981. *Philosophische Rundschau* 28:259–263.

T. M. Seebohm (deceased)
Mainz, Germany

© Springer Nature Switzerland AG 2020
T. Nenon (ed.), *Thomas Seebohm on the Foundations of the Sciences*, Contributions to Phenomenology 105,
https://doi.org/10.1007/978-3-030-23661-8_13

the well known fact that Husserl has a certain preference for the deduction in edition A[2] and that Fichte refers mostly to edition B will not be corroborated by further interpreting all the passages in both in which they refer to the deduction. What is at stake is a general *systemtheoretische* explanation of this and other facts.

A *caveat* is hence necessary with respect to the treatment of Fichte in Part II. What is of interest is the Fichte who contributed to the development of speculative *dialectic,* though he himself did not use this term to denote his method in the *Doctrine of Science.* Though he stresses the rigor of his deductions in the period before 1800, there are already passages at this time which indicate doubts concerning the possible fulfillment of the new methodological ideal (I 76/77, especially the footnotes, cf. I 5 15).[3] One could claim that such doubts determine his later development – just as similar doubts determine the development of the later Schelling. It can also be claimed that possible affinities between Fichte and Husserl are grounded precisely in these aspects of Fichte's philosophy. It must also be kept in mind, however, that precisely this aspect was the target of criticism by Hegel in 1802 and later, in the course of further development of speculative dialectic.

A similar *caveat* is necessary with respect to Husserl. For many the *Ideas I* is still the work which represents the final grounding of transcendental Phenomenology. However, as Husserl himself indicates, the problem of the final grounding is not solved here. Hence the final answer to the enigmas of Kant's transcendental deduction cannot be found here. In addition, the introduction of the transcendental ego in the *Ideas I,* a concept which seems to be its corner stone, was the target of criticism.[4] What the criticism claims is, in a nutshell, that the *Ideas I* either represent some subtle combination of transcendental psychologism and empirical idealism or else one has to presuppose the transcendental ego as a transcendental subject in the Kantian sense. At this point there is the temptation to compare Husserl's transcendental ego and Fichte's absolute I; and with this there arises the danger of a double equivocation. First the Absolute of speculative thinking refers to totality. It is rather doubtful whether Husserl, when he uses this term (Hua III, 198, 114f, cf. 140, Hua

---

[2]Cf. I. Kern, *Kant und Husserl* (Den Haag, 1964), 176ff.; and before my *Die Bedingungen der Möglichkeit der Transzendentalphilosophie* (Bonn, 1962), 181/182. Kern refers to material of Husserl concerning Fichte, loc. cit., 35ff., 292, 297.

[3]References to Fichte use *Johann Gottlieb Fichtes sämmtliche Werke,* ed. Immanuel Hermann Fichte (Berlin, 1845/1846). The translation *Science of Knowledge with First and Second Introductions* (1982, iibers. P. Heath and J. Lachs) has the page numbers of this edition in the margins. I refer to the *Wissenschaftslehre* with the translation *Doctrine of Science.*

[4]Cf. Natorp, P. 1901. Zur Frage der logischen Methode; mit Beziehung auf E. Husserls Prolegomenon zur reinen Logik. *Kant-Studien* VI:270–283; Natorp, P. 1917/1918. Husserls 'Ideen zu einer reinen Phanomenologie'. *Logos* VII:224–246; Kreis, F. 1930. *Phänomenologie und Kritizismus.* Tübingen; Zocher, R. 1932. *Husserls Phänomenologie und Schuppes Logik.* Mänchen. – The first phenomenological reaction with a turn against the transcendental ego was given by Gurwitsch, A. 1929. Phanomenologie der Thematik und des reinen Ich. *Psychologische Forschung* XII which is available in English in Gurwitsch, A. 1966. *Studies in Phenomenology and Psychology.* Northwestern University Press. About Husserl and the Neokantians cf. my loc. cit. note 2, Chapter IX and Iso Kern, loc. cit., note 2, II. Abteilung, which contains much more material.

I, 67)[5] has anything like that in mind. Secondly, as will be shown in detail in Part III, the transcendental ego as pole is neither "absolute" in the first nor in the second sense.

All comparisons beginning with certain parallels in concepts and terms in Husserl and Fichte must remain superficial if they are not made on the background of three other questions: (I) how is the realm of the transcendental determined in the beginning? (2) how is its accessibility for knowing determined? (3) what are the methods of explicating this realm? The enigmas of Kant's transcendental deduction point in this direction. The question, concerning how these enigmas can be solved – on the one hand in Fichte's speculative dialectic, and in Husserl's transcendental phenomenology on the other hand – can serve as a guiding thread for answers to these three questions.

13.1 The Enigmas of the Transcendental Deduction

Significant discoveries which occur in the history of the human intellect are discoveries of new viewpoints under which old puzzles and antinomies can be systematically ordered and solved. It is, however, also characteristic of such discoveries that they create other hitherto unknown sets of problems which *prima facie* seem to be *insolubilia* and provoke criticism. Two types of criticism can be distinguished. In the first type the *insolubilia* are identified and used in indirect arguments against the new viewpoint and serve thus at the same time as arguments for the philosophical position of the critic.[6] The second type of criticism tries to develop a higher standpoint of reflection from which (a) solutions for the insolubilia can be offered and (b) the old position is not simply rejected but restored and transformed on the new level. Fichte's and Husserl's critique of Kant is a critique of the second type. In order to analyse the differences in the solution they offer, it is necessary to exhibit systematically the insolubilia or enigmas of the idea of a transcendental deduction as formulated by Kant.

---

[5] All references to Husserl use the Husserliana (Hua), Edmund Husserl *Gesammelte Werke* (Den Haag, 1950), ff. The translations of the *Cartesian Meditations* (Hua I) by D. Cairns (Den Haag: Nijhoff, 1977); *Ideas Pertaining to a Pure Phenomenology and Phenomenological Philosophy* (Hua III) by F. Kersten (Den Haag: Nijhoff, 1983); *Formal and Transcendental Logic* (Hua XVII) by D. Cairns (Den Haag: Nijhoff, 1960) have page numbers in the margins following the Hua. Other material used like *Erste Philosophie I* and *II* (Hua VII, VIII); *Zur Phänomenologie des inneren Zeitbewusstseins* (Hua X); and *Analysen zur passiven Synthesis* (Hua XI) are not or only in parts translated. The Beilage of Hua VI, the Crisis, mentioned below is not translated either.

[6] Jacobi, F.H. 1787. David Hume über den Glauben oder Idealismus und Realismus. Breslau, in *Werke* (Leipzig, 1812–1825), Vol. II, 301ff. in his critique of the thing in itself is such a critic. In recent times analytical philosophers like J. Bennett, 1966. *Kant's Analytic.* Cambridge University Press, deal in a similar way with the problem of the "unity of transcendental apperception", i.e. after pointing out the problem the corresponding part of Kant's theory is rejected and the remainder is as far as possible reduced to the own position.

1. There is a fundamental criticsm which stands behind all other criticisms which are its specifications. It arises with the question: what are the warrants of the objective validity of the knowing which is displayed in the transcendental deduction itself? What is at stake here are neither objects of experience nor objects constructed in pure understanding. If the object would be an object of pure understanding then the objective validity of the knowing displayed in the transcendental deduction would be ruled out by the transcendental deduction itself. Kant's claim that such a knowledge does not pose any problems because reason is only concerned here with itself (B 23) does not really hold.[7] This becomes obvious if the specifications of the root question are taken into account.

2. Many interpreters have been tempted by the term itself to understand the "transcendental deduction" as a "transcendental proof" or a "transcendental argument." However, such an interpretation is, as I have shown elsewhere,[8] completely at odds with Kant's logical terminology. For Kant a deduction and also a transcendental deduction is a real exposition, *Sacherklarung, expositio realis* (AA VI, 248/249, cf. 229/230).[9] This explains why Kant, without violating his own terminology, can refer to what he called "expositions" in one place as "deduction" in other places (B 119, 121). What he says in (B 126/127) about his enterprise, namely to give an "analysis of the concept of a possible experience" in which an explication of its necessary factors and their interrelation is given, is hence the proper description of a "transcendental deduction." The only obstacle to such an interpretation of the term is that in # 20 and # 26 strings of sentences can be found, which have the character of chain syllogisms.[10] Closer consideration shows, however, that these syllogisms have the function of connecting that which is explained in the transcendental deduction, with the findings of the metaphysical deduction and the expositions of the transcendental aesthetics (B 143, 159/160).

According to Kant's Logic an exposition is an incomplete definition. A definition is the complete list of all characteristics, or markers, of a concept. He distinguishes furthermore between nominal and real definitions and expositions. Nominal definitions are complete if the given list of characteristics is sufficient to distinguish objects falling under a concept from all other objects. Real definitions and expositions have "to give its (the object's) inner determination by setting forth the possibility of the object out of its inner characteristics." (AA IX, 142–144).

[7] References to the *Critique of Pure Reason* are given as usual. References to the *Logic* and the *Metaphysical Elements of Justice* are given following the edition of the Prussian Academy of the Sciences (AA) (Berlin, 1902) f. listing the number of the volume and the pages.

[8] See my "Die Kantische Beweistheorie und die Beweise der Kritik der reinen Vernunft", *Akten des V.Internationalen Kant-Kongresses,* Vol. II (Bonn, 1982), 127–148.

[9] Cf., loc. cit., 139f.

[10] Cf. D. Henrich, 1969. The Proof-Structure of Kant's Transcendental Deductions. *The Review of Metaphysics* XXII:640–659. The criticisms of H. Wagner and H. Robinson in the *Kant-Studien* 1980, 352–366 and 1982, 140–148, should be taken into account.

It is a result of Kant's critical philosophy that only nominal definitions are possible with respect to objects of experience. Real definitions and expositions are only possible for mathematical objects and moral concepts (B 755 ff.). Furthermore he denies that philosophy is capable of giving definitions; it is restricted to expositions (B 758). But the "transcendental deduction" understood as "real exposition" is then possible only under the tacit assumption that the knowing displayed there has an access to the essence of experience; is capable of explicating "the possibility of its object out of its inner characteristics." Nowhere in the Critique a justification is given for the validity of this claim.

3. Kant's critics have pointed out that Kant promised a "deduction of categories" in the sense of demonstrating that the set of categories he offers is (a) complete and (b) necessary.[11] It can be shown that Kant explicitly denied that any further ground can be given for the *fact* that we have just this set of categories and no others (B 145/146). What he understood by completeness is no more than the claim that his list of categories allows us to derive all a priori principles of the natural sciences (B 110).

This criticism raised first by the German Idealists has however a deeper root in the transcendental deduction itself. Jacobi notes that behind the concepts of "affection," "giveness," and "passivity" there lurks the category of causality. He could well have realized that the same holds for the concept of "spontaneity." Other categories, used so to speak *per analogiam,* are used as well. That the concept of a "transcendental subject" implies some application of the category of substance is obvious. After introducing the concept of the "unity of transcendental apperception" Kant hastily adds that this "unity" is not the category of "unity" as applied to objects of experience (B 131). The concept of the "manifoldness" given in intuition is also an application of the categories of quantity *per analogiam.* It is likewise clear that the use in the deduction of modal concepts like "necessary" and "possible" cannot be understood in terms of the use justified in the postulates for empirical objects, but it can also not be reduced to pure logical necessity and possibility in Kant's sense.

If now there is an application of categories *per analogiam,* i.e. neither in an empty fashion to a noumenon, nor in the way in which they are applied to objects of experience, and if the transcendental deduction explicates the inner possibility of experience, then a "deduction" of the categories understood by Kant's critics must be given in order to justify the validity of the knowledge displayed in the

[11] Cf. J.G. Fichte, *Second Introduction to the Science of Knowledge,* I, p. 478; G.W.F.Hegel, *Science of Logic* (W.H. Johnstone and L.G. Struthers), Vol. I, 1929 (Allen & Unwin/Humanitas Pr.), 57/58; K. Reich, 1932. *Die Vollständigkeit der Kantschen Urteilstafel.* Berlin. Koerner, S. 1967. The Impossibility of Transcendental Deductions. *Monist* 51:317–331, has given a thorough critique of such an idea of deduction. It must be said, however, that – though many Kant interpreters believed in its possibility – Kant himself did not. On the other hand it holds, that Koerner did not explicitly deal with Fichte's and Hegel's attempt to develop such a deduction with the means of speculative dialectic.

"transcendental deduction." Since the "principles of logic" are implicitly applied in the deduction, they must likewise be "deduced."

4. Granting that Kant succeeds in showing that "things considered in themselves" and the noumenon occur only as indicators of the limits of knowing (B 310/311), the problem remains that "a possible experience" requires assuming a beyond, which is as such not an object of experience and is introduced in the transcendental analysis of a possible experience only by hypothetical assumption without being directly accessible in the analysis. The same holds for the synthetical unity of apperception and the spontaneity in or behind it, which has to be assumed to explain the analytical unity of consciousness which is given in the "I think" which can accompany all my thoughts (B 131/132). The question hence arises whether it is justifiable in a real exposition of the essence of experience to refer to factors, which do not only transcend the realm of objects of possible experience, but also the type of knowing which is given through the real exposition of possible experience itself.[12]
5. The difference between the analytic unity and the synthetical unity of apperception entails two concepts of self, the empirical self and the transcendental self (B 67f., B 152f., B 157f.). Because the transcendental self is as such inaccessible by definition, the problem concerning how the two are identical cannot be solved. Thus Kant's attempt to indicate the connection between both via the theory of the self-affection of the inner sense remains inadequate.[13] In addition, there arises a further question not even considered by Kant. It occurs only if the question of the status of transcendental knowledge in general is raised. It is the problem of the self of the transcendental knower, the "philosopher." This self cannot be identified with the empirical self, because it reflects the experience in which the empirical self is given. It cannot be identified with the transcendental self in the Kantian sense either, because the philosopher's self is accessible to him as the self which has transcendental knowing.

13.2 Fichte's Solution

Fichte introduces two methodical innovations: intellectual intuition and reflection applied to intellectual intuition. According to Kant intellectual intuition is impossible. It presents that type of knowing in which an *intellectus originarius* has its object, the noumenon in the positive sense. The concept of such a knowing is

[12] Within the first *Critique* the self as *causa noumenon,* cf. B 579ff., remains an empty metaphysical possibility. It receives content in Kant's practical philosophy, cf. V, 48ff.

[13] The problem of the "double world theorem" becomes most pressing in the *Religion within the Limits of Reason Alone,* XX VI, 31, 47. Cf. the English translation by Th.M. Greene and H.H. Hudson (Harper, 1960), 26/27 and 42/43.

completely empty and is possible as the result of abstracting from all positive characteristics of a finite intellect:

(a) Subject and object are not separated. For finite intellects objectivity is grounded in giveness. Nothing can be given to an infinite intellect.
(b) A finite intellect needs logical forms in order to combine *given* contents, hence in this intellect logical form and content are separated. Since nothing can be given for an infinite intellect, logical form and content are not separated in it.
(c) Since nothing is given in intuition to an infinite intellect it needs no forms of intuition, and that means, first of all, that the intellectual intuition of an infinite intellect is timeless (B 72, 158/159, 209, 212; cf. Fichte I 72).

Fichte claims to the contrary that the attempt to think the unity of transcendental apperception as pure identity and spontaneity *prior* to categorical form requires intellectual intuition. Such a thought must take place within the identity of the unity of transcendental apperception itself, and hence it follows that:

(a) Subject and object are not separated in this thought. Self-consciousness is pure identity.
(b) Logical form and content are the same, since nothing is given from the outside in this thought.
(c) The identity of the self is immediately intuited in that thought without something being given in a specific form of intuition. The intuition is, hence, intellectual and furthermore also atemporal (1458f., 463, 475f.).

Kant's rejection of intellectual intuition has thus to be reinterpreted. What it demonstrates is that the attempt to think the thing in itself as the noumenon in the positive sense is self-contradictory. The thing in itself is an impossible thing (*Unding).(l* 471/472).

The task of the *Wissenschaftslehre* is to deduce the emergence of all the factors which play a role in the transcendental deduction, i.e. the logical principles, the categories, and representation, ultimately grounding them and their theoretical functions in the practical. The first principle is the originating point which is grasped in intellectual intuition. Two series of development have to be distinguished. One is the series in which the factors emerge out of the pure acts of the absolute I. The other is the series in which the philosopher observes and reconstructs this emergence. The first is the existing system, the second is the system as represented by the philosopher. The philosopher is the empirical self who furnishes himself with the account of the emergence of his consciousness out of the absolute self (I 80, 454/455, 460). Since such an account is an account of and for an empirical self-consciousness, reflection can play a methodical role. "Reflection" has to be understood in the logic of the time. It is the determination of the content of a concept by listing its positive markers and is correlated with abstraction, i.e. the elimination of the negative markers which do not belong to the concept (IX, 94–95 (100)). If reflection is applied in intellectual intuition its structure will be transformed. The transformation leads to the specific method of the *Wissenschaftslehre*.

The logical aspects which occur in the beginning of the WL determine the basic structure of all further speculative dialectical thinking. Fixing first the principle of Identity "A=A" its transcendental condition is determined as *A is insofar A is posited by the I as identical I.* The formula under which the identity of the I has to be thought is gained by "instantiation": *I is insofar as I is posited by the I as identical I, I=I.*[14] In the first formula there is a difference between the being of A and the positing of A. In the second this difference collapses. Being and positing is the same. The I is its positing, it is pure positing *(Tathandlung).*[15]

The second principle is the principle of opposition. It is determined in its content, i.e. the "I," by the first principle. But in its form, the introduction of negation, it is unconditioned. This holds of the first series, the acts of the absolute I. Seen however from the viewpoint of a developed consciousness the principle is necessary in every respect, because in such a consciousness there is opposition, difference. The principle: *I op-posits the Non-I to the I* (I 101ff.) relates to the principle of Noncontradiction "-A not= A" in the following way. The principle requires that *I op-posits the Non-A to the A,* i.e. the principle of opposition as its presupposition. This in turn can be generated by instantiating A for I in the places which are opposited.

Neither the logical principles "A = A" and "-A not = A", nor their "transcendental" formulas contradict each other, but the transcendental formulas in which A is instantiated by I contradict each other. According to the first principle the I is nothing but its positing. Thus it is with the second in its entirety the I as well as the Non-1. The contradiction has, however, a pure logical aspect as well. Qua absolute I the I represents a unique highest genus. Nothing is outside it. Hence species can be generated out of it only, if it itself functions as specific difference and as its species at the same time. Given such a *dihairesis* – which is, of course, forbidden in formal logic – the negation of a genus is an analytical predicate of the genus and vice versa.[16]

In order to overcome the contradiction, a third act of the I is required: *In the I the I opposes the divisible I and the divisible Non-1.*[17] The analysis of the synthetic principle *divisibility* which harks back to the first principle as its ground, leads to the new contradiction: *The I posits the Non-I as limited by the I; The I posits the I as*

[14] Fichte is aware of this and other circularities in his "deductions" and considers them a belonging to the nature of reason. Cf. *Über den Begriff der Wissenschaftslehre,* I 62, 66ff., 72, 74, 78. The problem is closely connected with the undecidability of the dispute between idealism and realism, cf. I 429f., 435f.

[15] The meaning of *Tathandlung* can be explicated best by keeping in mind that *Tatsache, factum* in its original Latin sense is a juridical term. *Sache,* fact is considered as *Tatsache* to the extent in which it is the result of an act, *Handlung,* of an agent. The object A posited by the I is a *Tatsache.* Since in case of the positing of the I by the I the factum is itself an act, a Handlung, one has a *Tathandlung.*

[16] For a more detailed account of the logical aspect of developing dialectical contradictions and the type of formal logic presupposed here see my: The Grammar of Hegel's Dialectic, *Hegel-Studien* 11 (1976), 149–180.

[17] The translation of Heath and Lachs, "In the self I oppose a divisible not-self to the divisble self' (110) hides the dialectical character of Fichte's formula which I tried to preserve in the given translation.

limited by the Non-I, the latter being the bases of the theoretical part of the doctrine of science, the former the basis of the practical part.

With these steps not only the logical principles, but also the categories of quality: affirmation, negation and limitation, and, in the concept of divisibility, quantity and manifold ness are deduced in the specific significance they have with respect to an analysis of consciousness. The theoretical part develops the categories of relation, causality, substance and mutual determination, in increasing degrees of complexity until they are fit to explicate the power of imagination, then representation and with it time and space. In their more complex forms the stages represent at once different types of realism and idealism.

The method of the development is determined by the development of the first three principles. Given a synthetic term, its analysis, which harks back to previously developed content (at the most decisive points toward the end the first principle itself is involved), leads to a new contradiction. The analysis of the contradictory terms in turn determines the locution of a new synthetic term, which has to be understood as a new act of the I, generating a new category or a modification of it. The "observations" of the philosopher who takes his stand in intellectual intuition refer, hence, to his object, the first series, in a strictly formal way. It is formal in content because what is at stake is the explication of consciousness in terms of pure categories. It is formal in its method, because it is guided by the scheme produced in the first three principles.

The question, whether Fichte has fulfilled the methodical ideal of a closed system of speculative thinking which he himself has set up, cannot be discussed here. What can be claimed, however, is that already the first part of the Doctrine of Science offers specific answers to the enigmas of the transcendental deduction.

The answer to the first enigma concerning the status of transcendental knowledge is intellectual intuition. The answer to the second enigma is Fichte's "dialectic", in which the laws of reflection function in the specific way they have to function if they are "applied" to the content of intellectual intuition, i.e. the absolute I. The answer to the third enigma is given with the answer to the first to the extent in which the unity of transcendental apperception is concerned. Insofar as the thing in itself is concerned the enigma is answered by showing that the Non- I is not an independent entity, i.e. by overcoming, step by step, more and more refined types of realism, including critical realism. The answer to the fourth enigma is that the method of speculative dialectic allows us to deduce the logical principles of reflection and the categories as necessary factors in the emergence of consciousness. The answer to the fifth is the deduction of empirical consciousness out of the acts of the absolute I and its development into the I of the philosopher as the I that represents its emergence to itself.

Kant's shortcoming, according to Fichte is, that his method is regressive. Kant understands his highest principle as a hypothetically introduced condition and is not able to grasp it immediately as determing ground. However, a regressive method is necessarily based on unclarified grounds and hence is open to dogmatic misinterpretations and sceptical critique (I, 476/477).

13.3 Husserl's Solution

The root question concerning the transcendental deduction is: what are the warrants of the objective validity and what is the status of the knowledge displayed in the deduction. Fichte deals with this problem right at the beginning of his enterprise, while Husserl postpones this question. In most of his introductions to transcendental phenomenology it is acknowledged only in alluding to the necessity of a "critique of the critique," a further "apodictic reduction" and a "phenomenology of the phenomenological reduction" (FTL 294/295; Hua I, 67, 68, 178; Hua III, 197/198; Hua VII, 275, Hua VIII, 80, 169f., 378).

It can be said *a fortiori* that, in counterdistinction to Fichte's, Husserl's solution is based on a regressive method and furthermore, that the problem of the transcendental deduction is not solved by means of the transcendental phenomenological reduction. Given only the reduction and the type of analysis that is offered in *Ideas I,* a Fichtean critique is still possible.

The transcendental-phenomenological reduction can be understood *prima facie* as an explicit fixing of the attitude in which one is interested in a universal reflective critique rather than in conceiving objects in direct intention as in the mundane attitude. What.is of interest is not what is given in experience but rather the how of the giveness. The real existence of objects and of the world is neither denied nor even doubted (Hua III, 100 ff.; VII, 337). They are, on the contrary, presupposed as the correlates of my experiencing life and serve as transcendental guiding threads (Hua I, 87; Hua III, 368f.). The priority of the transcendental attitude is an epistemic, not an ontological priority. The giveness of the world and objects in it is neither adequate nor apodictic, at best there is a presumptive apodicticity with respect to the world. The giveness of my experiencing life is likewise not adequate, but it is given in an original evidence that is apodictic, (a) in its existence and (b) with respect to certain general structures, i.e. every critical reflection, even the possibility of such a reflection on such an evidence presupposes what is questioned as reflections on a higher level show. The epistemic priority of the transcendental attitude is also given, because it is in this attitude and not in the mundane attitude that the aforementioned specifications of evidence can be finally clarified. It cannot be denied that Husserl sometimes also claims an ontological priority of his transcendental *Seinsboden,* but this is (a) not *of* interest here and (b), as I think, misleading. The latter point cannot be further discussed in the present context (Hua I, 56; III, 115/116; VI, 410; VII, 410; VII, 338; VIII, 128/129, 311, 397).[18]

The realm of the "transcendental" is what is given in the stream of living experience united by inner temporality, a unity which can be grasped as unity without having it as a unity within a pregiven world. Included are not only the unity of the ego and its intentional acts but also the unity of the hyletic field. Excluded are the

[18] For a more detailed account of "apodictic evidence", its relation to original evidence and adequate evidence and the development of these concepts in Husserl see my *Die Bedingungen der Moglichkeit der Transzendentalphilosophie* (Bonn, 1962), 58ff.

transcendent unities of real as well as ideal objects, as the correlates of specific phases of the stream and its universal correlate, the world (Hua III, 9 ff.). The first task is the analysis of the mode of giveness of the transcendent in the transcendental or in the immanence.

The method of explication is eidetic description and analysis. Several points must be emphasized. Eidetic analysis is interested in onesided or mutual foundations between abstract factors, i.e. factors which do not represent self-sustaining entities, concrete wholes. Clusters of such factors are eidetic structures determined by "laws of essence" defined for foundation relations. A more detailed analysis would show that the so called morphological *eide,* i.e. material essences, always refer to structures which are still abstract and do not represent concrete wholes.[19] On the one hand this leads to certain restrictions on "eidetic intuition" with respect to material transcendent objects, but on the other hand it is important for the correct understanding of what can be reached if eidetic intuition is applied to structures of the transcendental sphere; or, in Kantian terms, it is used for the "analysis of a possible experience". Transcendent objects representing concrete wholes are always given on a certain background, in the last instance the background of the world. They are concrete wholes because they can be fixed in eidetic variation while varying the background. It could be shown that such a varying of the background is not completely free, i.e. there are cases in which a certain variation has an influence on the giveness of the concrete whole in question itself. What is significant at the present moment is that the unity of the stream of consciousness is the only concrete whole given in the transcendental attitude. No self-sustaining parts can be identified – only abstract factors and foundations which connect them.[20]

Neokantians have pointed out[21] that transcendental phenomenology if presented in the framework indicated is a brand of a subtle transcendental psychology because it presupposes the unity of my experiencing life in which eidetic intuition and self-objectification takes place. If there is an experience then an account of the unity of that experience must be given. Certain formulations of Husserl in the *Ideas I* seem to point in a similar direction. He refers to the ego as a transcendence in immanence (III, 138) and quotes Kant's "The 'I think' must be capable of accompanying all my presentations". Since in the original edition the restricting clause "whether also in Kant's sense I leave undecided" is still missing, the term transcendence in this passage can and has been understood as a reference to a factor in the immanence which transcends the sphere of immanence and constitutes its unity. How this factor is grasped would remain as unclarified in transcendental phenomenology as it is

[19] The III. Logical Investigation is, as Husserl himself emphasized, crucial for a proper understanding of his method. Cf. my "Reflexion and Totality in the Phenomenology of E. Husserl", *Journal of the British Society for Phenomenology* 4 (1973), 20–30.

[20] Given the IV. Cartesian Meditation neither the term "transcendental ego" as used in the beginning of the *Cartesian Meditations* nor the term "eidos transcendental ego" should be taken as referring to the ego as pole. The latter is an abstract momentum in the "ego in its concretion" i.e. transcendental subjectivity.

[21] Cf. note 4.

unclarified in Kant's transcendental deduction. The only way to answer it if one indeed wants to give an answer, would be in terms of a Fichtean intellectual intuition, i.e. the speculative thinking of German Idealism would remain, also with respect to transcendental phenomenology, the true *first philosophy.*

The other possibility is, of course, to emphasize what Husserl has emphasized later. "Immanence in the transcendence" is in itself a temporal difference, the difference between living presence and the part of the stream which can be brought to giveness only in acts of remembering. It is not the difference between the temporal and the atemporal and it is a property of the "ego" only because the ego is itself constituted temporally and is not a transcendent selfsustaining entity but an abstract factor like everything else in the stream.

Husserl's *prima facie* answer to a Neokantian interpretation of his conception of the transcendental can be found in his criticism of Kant's transcendental deduction. It has value to the extent in which it is a phenomenological and eidetic exposition of the structure of experience. It becomes problematic when Kant, facing certain difficulties, offers deductions, i.e. expositions in which entities are introduced by hypothetical substructions. In other words certain factors are assumed which are *ex definitione* not accessible for descriptive accounts. Such substructions are impossible for the phenomenological attitude. The whole question of the transcendental deduction, namely the problem of the objective validity of categories, is misconceived. A phenomenological analysis grounded in a transcendental aesthetics far more extensive than the one offered by Kant, can show how the categories emerge in experience together with the corresponding objects. Genetic phenomenology can show how logical categories are prefigured on aesthetic and prepredicative levels of experience and then emerge as ideal objects in intentional acts which imply idealisations and abstractions (Hua VII, 197ff., 231ff.).[22]

Taken for itself, such a criticism of Kant cannot count as a refutation of the criticism which can be raised by Kantian and then Fichtean viewpoints against transcendental phenomenology, because this criticism is directed against the very claim of phenomenology to be able to function as *transcendental* phenomenology and as first philosophy. What has to be explained is how phenomenological reflection is possible as transcendental reflection without explicating the unity of the realm of the phenomenological transcendental experience by means of the "substruction" of a transcendental subject that is accessible to only intellectual intuition.

The analysis of the stream of consciousness must be naive in the beginning, i.e. only after the surface structures of consciousness have been analysed can an analysis be given of the deeper structures and the possibility of phenomenological reflection itself. Seen from the viewpoint of the enigmas of the transcendental deduction and Fichte's attempt to answer them via the introduction of intellectual intuition, the central problem is the split between reflection and its object in which the unity of experiencing life is nevertheless itself given. It is that knowing in which subject and

[22] Hua XI, *Analysen zur passiven Synthesis* give the most corroborated version of this program. Husserl himself entitled one version of this lecture "Transzendentale Logik, Urkonstitution", cf. loc. cit. XIV.

object are one. The medium of Husserl's solution is his phenomenology of inner time consciousness. Three levels can be distinguished:

(a) The first level is the level of the constitution of the stream as reaching beyond the living present, of original selfgiveness. The constitution of this transcendence in the immanence (Hua I, 144/145; Hua XI, 204) requires specific intentional activities which occur as activities in the living present: recollecting, remembering and explicit expectations. What they grasp, however, is preconstituted in absolute passivity and what originally awakens them is passive association (Hua XI, 172f.).
(b) The next level is the structure of the passive constitution or inner time consciousness. Its form can be abstracted from inner time objects in the living presence as the form of the unity of primary impression, protention and the continuum of retentions in which content flows off. The form itself does not flow, it "stands." As objectified form it can only be grasped from flown off phases of content, which are reflected. The actual emergence of primary impression in the *nunc stans* of the actual Now cannot be an object of an explicit reflection, it is the place in which everything, hence also reflective activity, emerges. That this is the case in general can only be known indirectly by iterative reflection.
(c) The deepest level is the awareness of the unity of consciousness in the *nunc stans* itself. Nothing is grasped in such a consciousness. It accompanies all consciousness whether it is active or passive. It is the unity in which the actual now of reflection and flowing off phases which are the object of it stand together. Since it has no specific content for itself, it represents in awareness the concrete whole of consciousness and all that can be said about it is restricted to what can be said about it through the abstract factors which are already given on level (b). It is, however, a result of investigations on level (b) that the *nunc stans* of the actual Now and hence also the awareness of the concrete unity of consciousness can never be the object of an explicit objectifying reflection.[23]

In asking how solutions can now be given to the enigmas of the transcendental deduction it has to be kept in mind that the transcendental deduction has value from a phenomenological point of view only to the extent in which it is a phenomenological and eidetic exposition of the structure of experience. It has been shown above that the real meaning of the Kantian term "deduction" as real exposition indicates that such an evaluation does not destroy the actual intent of Kant's own enterprise. The answer to the first enigma can be developed out of the analysis of inner time consciousness. The *punctum saliens* is that the unity of time and also inner time is grounded in an atemporal spontaneity in Kant. For Husserl the location of that unity

[23]The question of objectification connects level b with level c. Most of the material relevant for this point is available in the C-manuscripts. Cf. loc. cit. in note 18, 126ff. and in note 20, 25ff. Concerning a controversy between methodological levels of the analysis and levels of constitution, see my review of Sokolowski, R. 1974. *Husserlian Meditations*. Northwestern University Press, in *Philosophische Rundschau* 23 (1976).

is itself connected with temporality. For Kant time and also inner time is though in terms of an undimensionally stretched "line" indefinitely open at both ends. In Husserl inner time is a structure which has its origin in the *nunc stans* of the absolute Now in which all content emerges and from which all content flows off. It is at this point that the unity of inner temporality is constituted in absolute *passivity*. This model allows us to think the unity as Kant thinks it as not explicitly objectifiable *in concreto* but, nevertheless, as connected with the self-experience and temporal self-experience as it is given for reflection. The possibility of reflection and also transcendental reflection is explicable in terms of this temporality. Reflecting and reflected stand together in the unity of time consciousness, but it is also the structute of this consciousness as temporal that makes their split possible. A reflection grasps the reflected as a content which belongs, already in the living presence, to the flowing off phase of this present. The reflection itself emerges in the actual Now. That this is the case can be shown on level (b) in iterative reflections.

The answer to the second enigma is solved in the clarification which eidetic intuition can provide especially with respect to an analysis of the structure of consciousness. The answer to the third is given in the rejection of all "deductions" and "transcendental arguments" for the objective validity of categories as well as for a certain set of categories as the only possible set. Static and then genetic analysis must first clarify the emergence of categories in strict correlation to the emergence of the objects corresponding to them, and then their giveness as ideal objects on the level of formal ontology which is based in acts of abstraction and a specific eidetic variation. It can be said that this solution presupposes a certain Leibnizean correction of Kant's sharp distinction between sensibility and understanding. For Husserl it is the schematized category which has priority, and the pure category is developed out of it. With respect to its subjective aspect the fourth enigma is solved in the same way in which the first is solved. With respect to the objective aspect an answer can be provided on two levels. First, the hyletic field belongs to the sphere of immanence and that immanence is also opposed to a transcendence in the immanence. There is no affection from the outside. Secondly must be kept in mind that the difference between the mundane and the transcendental attitude is an epistemic difference, i.e. no ontological claims are made about subject and object or their interactions. Consciousness is one and the same "reality" seen in different attitudes. It should be noted that the epistemic priority of the transcendental attitude presupposes that the mundane attitude has genetic priority, i.e. one can never enter the transcendental attitude without having been in the mundane attitude before, but one can live in the mundane attitude without ever entering the transcendental attitude. The fifth enigma can be solved in pointing out that both the mundane and the transcendental attitude – and hence the giveness of the subject as belonging to the mundane and to the transcendental realms – are ultimately both grounded in the unity of time consciousness, i.e. the mundane subject and the transcendental subject are the same. It is the self of the phenomenologist which, after entering the transcendental attitude and the critique of the critique, becomes aware of this identity.

13.4 Comparison

The *tertium comparationis* in Fichte's and Husserl's solution cannot be the *ego*. In Husserl the ego is (a) an abstract factor of consciousness and there are (b) reduced types of consciousness, e.g. the primordial ego in the fifth Cartesian Meditation which would not be called "ego" in philosophy and psychology. They correspond to subhuman personalities. Though in the beginning of the practical part of Fichte's *Doctrine of Science* we are confronted with similar stages in the development of the I it holds nevertheless that the Absolute is an absolute I and that time which occurs together with the power of imagination is derived in a logico-dialectical process. What corresponds to Fichte's absolute I is in Husserl the *nunc stans* of the third level of time constitution.

Prima facie the difference between both is that Fichte's approach is an approach in which the unity of transcendental apperception is as in Kant understood as an atemporal entity. Husserl, on the contrary, provides us with an analysis of the "inner sense" which shows that what corresponds to the unity of transcendental apperception can be understood as the origin, *Urquellpunkt,* of temporality which is as such connected with temporality and not located in some beyond.

The unity of transcendental apperception in Kant is that which explains the analytical unity of consciousness, it is that synthetical unity of consciousness which is prior to all analytical unity (B 133). The *tertium comparationis* must hence be the question: what is the account that is given by Fichte and Husserl for the unity of consciousness?

The formal mode employed by Fichte is dialectical, i.e., what is grasped as absolute is thought in explicating it through the forms of the principles of logic. More precisely these principles emerge in the beginning from the acts of the absolute I. Since it is an absolute medium that is explicated, an immediate unity, the second act, which introduces difference, introduces difference as contradiction, which requires synthesis. It is characteristic of Fichte's dialectic – and Hegel was critical of this – that synthetical terms are terms which do not explicate the contradiction as such, but are terms which are posited in between the opposites. The first synthesis term is limitation. All others are reached in a "downgrading" in which (a) the interpretation of the synthesis term through the opposites synthetized by it leads to two new concepts standing in contradictory opposition and (b) a new limiting synthesis term is found for the newly developed contradiction. Seen formally, the result of the theoretical part of the *Doctrine of Science* is hence complete externalisation via infinite limitation, the infinite "in between" of the power of imagination and with it, time. Characteristic for Fichte's positive dialectic, i.e. the method of establishing synthesis terms, is that they are "found". The philosopher as observer finds them in consciousness as that which makes possible the unity of consciousness in its oppositions. It is this "finding" which is responsible for the *quasi* descriptive character of Fichte's dialectic.

Consciousness is a totality, i.e. a unity which has nothing outside it which limits it. The tool of the explication of this totality is the grammar of the logic of the concept. That form and content are not separable in this procedure holds equally for the "observing" philosopher as for the acts of the absolute I. The form of the explication is generated in the acts themselves in which the content occurs. The content, in the beginning, is nothing else than the principles of logical form and the logical categories. This is made clear in the first three paragraphs of the *Doctrine of Science* but also in the examples referring to objective applications in section E III 2b. It is Fichte's discovery – which grounds the further development of the dialectic German Idealism – that the application of the grammar of the logic of concepts to the analysis of totality, i.e. a unity for which no principle of differentiation from outside is possible, leads (a) necessarily to dialectical contradictions in which the negation is determinate negation and (b) offers therefore the possibility of mediating terms (1, 110–123).

What is the "grammar" of Husserl's account? That which holds for descriptive analyses in general has been mentioned above. The "grammar" is the logic of the whole and the parts as explicated in the *Third Logical Investigation.* Consciousness is a concrete whole to the extent in which it is grounded in temporality. An explicit thinking of temporality in which something is explicated as "observed" is possible only on level (b). The formal viewpoints used by Husserl in the description – in addition to those taken from his theory of the whole and the parts – give hints about the nature of the grammar which is operative here. He speaks about the continuum of retentions, continua of continua, multidimensional continua and so forth. There can be only one answer to the question of how these continua can be thought as connected to the *nunc stans* of level (c), that is the actual Now in terms of Husserl: The actual Now has to be thought as the transfinite of every possible series of objectifying reflections on series of retentions. That is, it is that new element which is an expression for the whole of the first chosen set of retentions in its natural succession according to the law of its succession. It is possible to think it as the new "number" as the limit towards which all the numbers given to the retentions of the first succession stretch and which is in "equal" distance to all of them. This means, of course, that a real reproduction of the succession of the temporality as a whole up to the actual Now is impossible (Hua X, 196/197). A more detailed account cannot be given here.[24] It should be clear, however, that the grammar in which Husserl thinks the whole of consciousness is the grammar of mathematics, more precisely the grammar of post-Cantor mathematics. This should not come as a surprise. If Husserl's solution to the enigmas of transcendental deduction is grounded in a reinterpretation of the structure of the "inner sense", and if it is the inner sense in Kant that provides the a priori intuition for arithmetic then a reinterpretation of the inner sense must involve a reinterpretation of number.

[24] Cf. Cantor, 1962. *Gesammelte Abhandlungen mathematischen und philosophischen Inhaltes,* ed. Zermelo. Hildesheim, 195, cf. also 168, 390.

Traditional philosophy, including Kant and Fichte, knew only the opposition of the finite and the actual infinite with respect to mathematics. To think the actual infinite was considered to be a task of speculative thinking and it was the logic of the concept which served as the tool for thinking totality. The "bad infinity" of mathematics, understood as an "in indefinitum" was finally banned by Hegel in favour of a dialectical-speculative concept of infinity.

The introduction of the transfinite and hierarchies of the transfinite between the finite and the actual infinite has opened up new possibilities of thinking "wholeness," corresponding also with respect to consciousness. It can be said that Husserl's solution of the problems connected with Kant's transcendental deduction and with transcendental reflection in general stand in the wake of the discovery of the transfinite. Fichte's solution, however, remains in principle in the medium in which totality is thought in the framework of the logic of concepts; a logic which, already in Kant, was strictly separated from extensionality in general. Yet it must be said – and this is the formal basis, in which the affinity between Fichte and Husserl is grounded – that Fichte's dialectic is closer to the logic of mathematics than is dialectic in its purified form in Hegel. For Fichte the third principle leads via limitation to divisibility, i.e. the capacity for quantity in general (I, I 08). To reiterate, the dialectically derived extreme of the capacity for quantity is the power of imagination for which there is time (1, 217). Though this affinity of Fichte's dialectic to "external relation" has been criticized from the viewpoint of purified speculative thinking, it holds nevertheless that everything Fichte says concerning time is said from the standpoint of the grammar of the logic of concept and the unmediated opposition of the finite and the actual infinite.

Fichte claimed – and Hegel gave him full credit for this – that his philosophy represents the spirit, not the letter of Kant's philosophy. Yet one could claim that the recognition of a supersensible world which frames the world of experience on both sides is exactly that which goes against the spirit of the Critique and that the attempts to think this *mundus intelligibilis* with the aid of speculative dialectics goes against the real spirit of Kant's philosophy. This spirit expresses itself in its regressive method, i.e. precisely that which Fichte criticizes. The problem of Kant's philosophy is that at the point in which the regress stopped it presupposed an atemporal entity behind the unit of transcendental apperception. The spirit of modern metaphysics (i.e. the idea of a method in which all eternal truths of reason can be deduced from a fixed starting point – whether it be the absolute I of Fichte or Being as presuppositionless topic does not matter) could be restored if a method was found to explicate and decipher this "absolute" behind critical reflexion. Husserl's transcendental phenomenology is a solution for the root problem of the transcendental deduction in which (a) the highest unity becomes an immanent factor of consciousness itself in the sense of the transfinite, not the actual infinite and (b) no limits are set up for critical regress which can stretch towards the transfinite without ever reaching it yet knowing it as its telos.

Critical philosophy though apodictically grounded, becomes an infinite task. Husserl's transcendental phenomenology is hence much closer to Kant's method and hence, as critical transcendental philosophy, it exemplifies the spirit of Kant's method to the extent in which this spirit is opposed to the spirit of modern metaphysics since Descartes. Fichte's philosophy, however, must count as a revival of the spirit of Cartesian metaphysics on a higher level, the attempt of establishing it once again with the help of new methodical tools beyond Kant's Critique.

Thomas M. Seebohm was professor emeritus at the Johannes-Gutenberg Universität Mainz. For a more detailed description of his career and his work, please consult the "Introduction" to this volume.

Chapter 14
The Paradox of Subjectivity and the Idea of Ultimate Grounding in Husserl and Heidegger

Thomas M. Seebohm

Abstract This chapter reviews J.N. Mohanty's analysis of Jaakko Hintikka's attempt to reformulate Husserl's notion of intentionality as intensionality conceived along the lines of analytical philosophy's discussion of possible worlds theory. It shows that the outcome of applying viewpoints taken from a phenomenological semantics on the level of predication to problems connected with the "possible worlds semantics" as considered in the analytical tradition was that the "world" as it is to be understood in the context of formalized logic cannot be understood as some kind of concrete world, the real world in which we live, the "worlds" in quasi -positionality or worlds constructed in metaphysical speculations as correlates of other intellects. Phenomenological reflection shows how they are the result of a peculiar abstraction and idealization which separates them in principle from all other conceptions of world. Thus, the paradoxes which are connected with the attempts to connect the different concepts of "world" vanish when reflections on the genetic origins of these kinds of abstractions are undertaken.

This chapter argues for a reading of transcendental phenomenology in Husserl under which the objections of the early Heidegger (and others) are rendered pointless. It suggests, however, that there is a positive claim in Heidegger's *Grundprobleme* that reveals a different critical point. It is impossible, according to Husserl and his reflections, to achieve ultimate grounding in the natural attitude, i.e., under the part of the paradox that explicates the being of subjectivity as being in the world. Heidegger apparently thought otherwise, at least in the beginning. of his career. The final section of this chapter shows that Heidegger's *Denkweg* ultimately leads

This chapter was previously published in the following book:
The paradox of subjectivity and the idea of ultimate grounding in Husserl and Heidegger, in Phenomenology and Indian Philosophy, ed. Chattopadhyaya, D.P. et. al., pp 153–168 (SUNY Press, 1992)

T. M. Seebohm (deceased)
Mainz, Germany

© Springer Nature Switzerland AG 2020

T. Nenon (ed.), *Thomas Seebohm on the Foundations of the Sciences*,
Contributions to Phenomenology 105,
https://doi.org/10.1007/978-3-030-23661-8_14

precisely to the result which can be foreseen from Husserl's point of view, but is rather part of Heidegger's later universal criticism of transcendental philosophy.

This chapter makes the case that this is precisely why Husserl bracketed the natural attitude and why an ontological interpretation of the reduction destroys the possibility of reaching the goal that Husserl declared to be this goal: to find a meaningful explication of ultimate grounding. This also why Heidegger had to abandon the original program of the *Grundprobleme* that was connected with his criticism of Husserl, namely, to fulfill Husserl's program of ultimate grounding in asking first the question of the being of the subject and then the question of being.

The paragraphs preceding the explication of the interplay of psychology and transcendental phenomenology in the *Crisis*[1] develop the formula that, according to Husserl, elucidates the principle of the interplay. It is the formula of the paradox of subjectivity. This formula is supposed to explicate the root problem with which the development of Husserl's phenomenology was infected from its very beginning. It is, however, also the principle of the criticism that has been raised by Martin Heidegger against Husserl's transcendental phenomenology. Before dealing with this criticism it is useful to elucidate the formula of the paradox of subjectivity with the aid of some reminders concerning the problem of psychologism in Husserl.

Husserl's arguments against psychologism[2] in the "Prolegomena" tell us that psychologism is simply a kind of relativism and that relativism can be refuted because it is self-contradictory. It has been said that the argument is spurious because the argument against relativism is spurious.[3] Indeed, the argument was later restated by Husserl himself.[4] According to psychologism, empirical psychology as a science is the final arbitrator in all questions concerning cognitive validity claims. Husserl argued that psychologism implies relativism. Relativism denies that scientific knowledge has universal objective validity. Hence psychologism denies the objective validity of its own judgments concerning cognitive validity claims. Furthermore, Husserl had pointed out already in the "Prolegomena" that there is one type of relativism which cannot be refuted, viz., the individual relativism which claims the relativity of validity claims not for the species but for each individual.[5] In connection with Husserl's later critique of Dilthey and worldview philosophy another type of relativism also entered in, namely historism.[6] Husserl did not offer

---

[1] E. Husserl, *Die Krisis der Europäischen Wissenschaften und die transzendentale Phänomenologie, Husserliana* VI, 1954, §§53, 55 (Henceforth *KR*).

[2] E. Husserl, Prolegomena, *Logische Untersuchungen* I, *Husserliana* XVIII, 1975, §§3438 (Henceforth *LU*).

[3] D. Føllesdal, *Husserl und Frege*, Oslo, 1958. Cf. J.N. Mohanty, *Husserl and Frege*, Indiana U.P., 1982 20 ff.

[4] E. Husserl, "Philosophy as a Rigorous Science, in: *Phenomenology and the Crisis of Philosophy*, Harper & Row, 1965, esp. 92–105.

[5] Prolegomena, §35.

[6] "Philosophy as a Rigorous Science," 129—Lauer and others translate "Historismus" as "historicism". I follow the terminology of K. Popper: *The Poverty of Historicism*, Routledge & Kegan Paul, 1960 in distinguishing between historism and historicism. Historicism is represented by theories—like those of Hegel and Marx—which attempt to give an account of universal history as

arguments against historism. He only emphasized its devastating consequences. It is obvious that consequentialist arguments cannot be directed against historism. Like individual relativism and unlike psychologism, historism is not committed to the belief in the objective validity of scientific knowledge. Contrary to the mistaken formulations of the "Prolegomena," there is, hence, in Husserl, no argument against relativism possible in a mundane phenomenology. There is an argument that points out that the claim of empirical science to function as final arbitrator in questions of epistemology leads to relativism and that relativism contradicts the validity claims of science.

In the later writings Husserl considered the transcendental-phenomenological reduction as the only remedy against psychologism and anthropologism in the broader sense, i.e., the sense that includes historism and other kinds of relativism. This broader sense includes even the phenomenological research offered by the *Logical Investigations*. They are characterized in the first edition as "descriptive psychology." Psychologists of the time frequently claimed that descriptive psychology is the basic tool of epistemology.[7] Already in 1903 Husserl rejected this view.[8] Although it was not made clear enough in the first edition, he claimed that the descriptions in the *Logical Investigations* do not refer to empirical persons and psychic facts. Rather as phenomenological they refer to the a priori essential structures of the experience in which logical objects are given. A description of inner experience and structures of intentionality is thus not an empirical psychology, although it can still be understood as a naturalistic discipline. Twenty years later[9] Husserl wrote that the *Logical Investigations*, though not an empirical descriptive psychology still represents a subtle psychologism. Indeed, any attempt to justify validity claims by means of a turn to the subject without explicitly performing the transcendental phenomenological reduction represents a transcendental psychologism of some sort.[10] If only the eidetic reduction were performed in investigations of structures of experience, then the investigation would not be phenomenology proper but rather eidetic psychology. Since in this case there would be no transcendental phenomenological reduction, the being of the subject would still be a being in the world. Thus, eidetic psychology is later characterized as a subtle type of transcendental psychologism or even anthropologism.

The relation of the tendencies of descriptive and then eidetic psychology towards either transcendental psychologism or transcendental phenomenology is by no means a negative relation of mutual exclusion. On the contrary, Husserl in his later

---

a whole. Such an account is seen from the viewpoint of historism, which represents a kind of relativism, as nothing else than bad metaphysics.

[7] Cf. Spiegelberg, H. *The Phenomenological Movement*, Vol. I, Nijhoff, 1960, esp. 57 (about Stumpf).

[8] Cf. the Preface to the Prolegomena, *Logical Investigations* Vol. I. (in the translation of Findlay, Routledge & Kegan Paul, 1970, 47.)

[9] *Formal and Transcendental Logic*, Husserliana XVII, §§56, 99 (Henceforth *FTL*).

[10]. For an account with more references, see Th.M. Seebohm *Die Bedingungen der Möglichkeit der Transzendentalphilosophie*, Bonn, 1962, 156–159.

writings asserts again and again that all the contents of eidetic psychology's descriptions are also without restriction valid in transcendental phenomenology. The only difference is the attitude with which these contents are considered. In eidetic psychology they are still considered in the mundane attitude. In transcendental phenomenology they are considered in the transcendental attitude that is the result of the reduction. Furthermore, in addition to the Cartesian path there is a path to the transcendental phenomenological reduction via psychology. All that is necessary there is a thorough critical phenomenological reflection on the presuppositions of the descriptions of eidetic psychology itself, i.e., an intentional analysis of the cognitive activities of eidetic psychology itself.

The paradox of subjectivity which explicates the interplay can be now formulated as follows: On the one hand, the world and all its contents are given as the correlate of consciousness. On the other hand, we can think of the being of consciousness only as a being in the world. The second part of the paradox explains why anthropologism is a necessary part of the self-apprehension of subjectivity. The "where" of its being can be apprehended only in the world. This, in turn explains why such a self-apprehension is necessarily infected with relativism. To be in the world is to be in one specific context in the world. Whatever is in the world in a context is in its being determined by this context. But under this presupposition it must be admitted that the way in which consciousness has the world as its correlate it determined by the specific context and, therefore, is different in different contexts. The additional puzzle is that the way in which this determination is thought depends upon the basic structure of the worldview and the different ways in which determinations can be thought in different. worldviews.

The paradox of subjectivity is helpful in some respects but leads to some other difficulties in the theory of the reduction. Given the paradox, the reduction can be understood as a decision in favor of the first part of the paradox. It is not obvious in the beginning that a full account of the paradox can be given after the reduction, and thus it is not obvious that the overcoming of relativism is again anything more than the decision to deny the motives for accepting relativistic positions.[11] A new evaluation of the controversy between Husserl and Heidegger is possible. The existence of this controversy has been neglected in the literature, last but not least because there are very few sources and most of them have been published only recently. There is, in addition, an interesting historical question. The "paradox of subjectivity" in the *Crisis*, where the being of subjectivity is at least under one perspective characterized as a "being in the world," was developed by Husserl after he had read the early works of Heidegger. It can be asked, therefore, whether he was influenced by Heidegger in this respect and whether the paradox can be considered as Husserl's own answer to Heidegger. The following considerations are, however, systematic considerations. The historical and biographical aspect is not of significance here.

[11] For a criticism in this kind, cf. G. Funke, "Practical Reason in Kant and Husserl", in: *Kant and Phenomenology*, Center for Advanced Research in Phenomenology and U.P. of America, 1984, 129.

Husserl tells us in his critical remarks about Heidegger that though Heidegger offers an intentional psychology, he rejects the reduction and therefore does not have a real understanding of the constitution of the object and of reality. His philosophy, like the philosophy of his forerunners Dilthey and Scheler, is thus an anthropologism connected to an objectivism and naturalism. In counter-distinction to the use of this term earlier, e.g. in "Philosophy as a Rigorous Science." "Naturalism" here signifies the presupposition of the mundane attitude in which the existence of the world is not bracketed because the reduction is not performed.[12] In a nutshell, this criticism says nothing else but that Heidegger has chosen the priority of the mundane self apprehension of the subject, i.e., he has chosen the part of the paradox which Husserl considered to be secondary in a sense that needs some clarification.

Heidegger's early conception of phenomenology presented in the lecture course *Grundprobleme der Phänomenologie*[13] confirms this thesis and answers in addition the question why Heidegger preferred the other part of the paradox: Phenomenology is the method of scientific philosophy and must be distinguished from worldview philosophy.[14] The method of scientific philosophy includes the methodical ideal of ultimate grounding (*Letztbegründung*). The development of the idea of ultimate grounding itself is the driving force of Heidegger's criticism of the transcendental-phenomenological reduction and Husserl's transcendental phenomenology in general.[15] Husserl speaks in his theory of the reduction about an absolute and apodictic being of the subject and a being of the world that is only a being relative to this absolute being of the subject. What is presupposed here, not asked and not answered, is the question of being. If one asks, however, the question of the "being of subjectivity," then it can only be understood as "being in the world." This understanding, however, cancels the transcendental-phenomenological reduction. All said historical questions set aside, it can be said from a systematic point of view that Heidegger rams the paradox of subjectivity against Husserl and his idea of reduction at this point.

Together with this criticism Heidegger proposes a new type of reduction which shall serve the idea of ultimate grounding better than Husserl's reduction. At this point he still calls it "phenomenological reduction."[16] It is the reduction of the beings (*Seiendes*) to being (*Sein*) in the ontological difference. The ultimate grounding is, hence, an ontological grounding. This idea needs some explications in Order to arrive at a proper comparison with Husserl's idea of ultimate grounding. Such an explication can also explain why Heidegger in the subsequent development first drops the term "phenomenology" and later speaks first about "hermeneutic phe-

---

[12] Husserl's notes about Heidegger are published in Diemer, A. *Edmund Husserl, Versuch einer systematischen Darstellung seiner Phänomenologie*,,Meisenheim, 1956, 2930. See also **FTL** §§56, 99.

[13] *Gesamtausgabe*, vol 24, Frankfurt Klostermann, 1975, (Henceforth *GP*).

[14] *GP* 1–3.

[15] *GP* 255 ff.

[16] *GP* 26, 31.

nomenology," then about "hermeneutics" and finally abandons even this term as characterizing his new way of thinking.

Joseph Kockelmans[17] has shown that the *Grundprobleme* are heavily influenced by Heidegger's interpretation, critique and reformulation of Kant's doctrine of method. The lecture *Phänomenologische Interpretation von Kants Kritik der reinen Vernunft*[18] was given immediately after the *Grundprobleme*. Heidegger's early methodological reflections are, hence, the fruit of his critical considerations of both Husserl and Kant. The significance of the philosophers of classical antiquity and the middle ages shall not be denied with this thesis. Systematically considered they have, however, significance for Heidegger because he thought he succeeded in his criticisms of modem philosophy and first of all of the philosophies of Kant and Husserl.

Heidegger's ontological idea of ultimate grounding in the *Grundprobleme* is critical to the extent in which it is grounded in the critical distinction between beings and being. It is transcendental because it transcends the realm of beings towards being.[19] The ontological structure of existence (Dasein) is grounded in the transcendence of being.[20] The a priori is understood as the form of the pre-ontological understanding of being as beings. An analysis of the a priori—i.e., Kant's enterprise—under which beings are given is, hence, in the final instance only possible from the viewpoint of the ontological difference, Heidegger's "phenomenological reduction" in the *Grundprobleme*.

The original intention of Heidegger is, therefore, to replace Kant's transcendental logic with a transcendental ontology,[21] more precisely to show that a transcendental ontology is the ultimate ground for a transcendental logic. At the same time a transcendental ontology is supposed to be the ultimate ground also for phenomenology and its theory of the a priori.

Some of the difficulties of Heidegger's idea of an ontological interpretation of the idea of ultimate grounding, which will be discussed in detail later can be foreseen already from a Kantian perspective. The Kantian ultimate conditions—it is Fichte who first demanded that they ought to be understood as ultimate determining grounds—are for Kant "transcendent" in the sense of "absent." Neither the subject nor. the object can be known "in themselves." Heidegger's being has the same character with respect to pre-ontological understanding. Ontological understanding still has this character to the extent which being cannot be thematized "in itself" but only in the ontological difference. After abandoning the ideal of ultimate grounding, Heidegger developed a new criticism of Husserl. He says that Husserl's phenome-

---

[17] For the following cf. J. Kockelmans, "Kant's Method and Heidegger", in: *Kant and Phenomenology*, Th. M Seebohm & J.J. Kockelmans eds, Center of Advanced Research in Phenomenology & UP and America, 198, 161–183,—see also Kockelmans, *Heidegger and Science*, Center of Advanced Research & UP of America, Washington DC, 1985, 53 f.

[18] *Gesamtausgabe*, Bd. 25, Frankfurt, Klostermann, 1977.

[19] *GP* 20 f. Cf. Kockelmans, *Kant's Method,* 165 f.

[20] Kockelmans, *Kant's Method*, 1997.

[21] Kockelmans, *Kant's Method*, 180.

nology is grounded in a "metaphysics of the present."[22] It is grounded in the evidence of what is originally given and thus it fails to recognize the priority of the transcendent, the absent, namely being.

This criticism has been corroborated later by Derrida.[23] It reveals, however, also the difficulties of the possibility of the idea of an ontologically ultimate grounding of the Grundprobleme. Being as ultimate ground cannot be "assumed" like a Neokantian hypothesis. An ultimate ground cannot be represented after ultimate grounding is achieved as something which is not there in some "*aletheia*," but "absent in principle." But if so, then Heidegger himself will still represent some "metaphysics of the present." He can be defended against such charges by pointing out that in the end he himself abandoned the idea of "ultimate grounding." This will be explained later. It is clear, however, that if he himself abandons the idea of an ultimate ontological grounding, then his early criticism of Husserl is pointless. He himself could not deliver what he promised to deliver.

Before looking into this matter, we have now first to investigate how Husserl's idea of ultimate grounding by choosing the other part of the paradox may prove its mettle against the charges raised by Heidegger. In order to achieve this goal, it will, however, be necessary to reject some essential elements of Husserl's own self-understanding. It will be necessary to remove everything which implies ontological theses and, hence, also ontological idealism.

Husserl himself quite frequently mentioned that critical problems remain after the transcendental phenomenological reduction. If transcendental phenomenology would be guided only by a naive trust in the newly discovered apodicticity without really knowing the nature of this apodicticity, then these problems would indeed remain unsolved. It is neither necessary nor possible to give a full account here of this "critique of the critique." What is at stake is the paradox of subjectivity and the problem that it creates for transcendental subjectivity because, according to Heidegger, the question of being is not asked and the nature of transcendence is not grasped properly. In terms of the paradox, this could be done only by giving priority to the other part of the paradox. In the "critique of the critique" there are two places in which the problem of transcendence surfaces again: (1) the problem of the self-givenness of the subject in the constitution of inner temporality and (2) the problem of "solipsism," which is actually the problem of the givenness of the other, the alter ego.

With respect to the first problem, three aspects can be distinguished: (a) The unity of consciousness is taken for granted in the analyses following the transcendental phenomenological reduction.[24] Thus, the phenomenology of inner time consciousness and its constitution is an essential part of the "critique of the critique."

---

[22] "Die Onto-Theologische Verfassung der Metaphysik" in: *Identität und Differenz*, Pfullingen, Neske, 1957.

[23] Derrida, J., *La Voix et le Phenomene*, Presses Universitaires de France, 1967

[24] *Ideen zu einer reinen Pḧäneomenologie und phänomenologischen Philosophie, Buch I. Husserliana* III, 1950, 197 f, 209, (Henceforth *ID I*). *Cartesiansische Meditationen und Pariser Vorträge, Husserliana* I, 1950, 81, 109 (Henceforth *CM*)

(b) The transcendental phenomenological attitude requires a split between observing and observed consciousness.[25] How such a split is possible in the unity of consciousness can only be understood if the temporal character of consciousness is properly analyzed. (c) The apodictic self-givenness is original selfgivenness but not an adequate self-givenness and it is restricted in its apodicticity to the present phase of inner temporality.[26] The analysis of inner temporality is, therefore, also of significance for a critique of the proper meaning and the scope of apodictic evidence.

A brief outline of the formal structure of the constitution of time consciousness indicates the answers that can be given to these questions from Husserl's own phenomenological research. The original temporal flux is constituted on the lowest level of passive constitution.[27] The abstract moments that constitute the whole of the flux are: (i) the moments that constitute the form of the flux, i.e., the actual now or *nunc stans* with its protentional horizon and the continuum of retentions in the retentional horizon, and (ii) the contents of the flux, which emerge as primary impressions in the actual now and flow off from there into the continuum of retentions. From the present temporal perspective, the contents vanish into the retentional continuum, i.e., they flow off into a temporal realm which is not given in original evidence.[28]

An explicit consciousness of inner temporality is constituted only on the level of intentional acts of remembering, expectation and immediate reflections on the flux of the living present itself. All these acts occur within the temporality that is constituted on the lowest level—like everything else in consciousness, including the ego. They are themselves temporal and constitute in their original temporality a consciousness of the "stream of consciousness" that stretches beyond the living present into a past that is not originally given. They constitute the consciousness of duration and succession and with it the consciousness of what is absent from the present.[29]

We can emphasize three points:

(1) The formal structure of the original flux can be recognized only because it is to be found in every flowed off phase of time that can be an object of reflection. The *nunc stans* of the actual now in the living present cannot, however, be objectified by reflection. Rather, it is the place in which all objectification occurs. The split that occurs between observing and observed consciousness can thus be explained by the temporal structure of consciousness, which is pregiven for all acts of reflection.

(2) Inner temporality and its unity are not the result of synthetic activity of the ego on active consciousness. On the contrary, all these activities occur in this medium, which is pregiven for them in "passive" constitution. There are classes of intentional acts in which consciousness and its temporality is given for itself

---

[25] *CM* 73; *Erste Philosophie II*, *Husserliana* VIII, 1959, 440–441, (Henceforth *EP II*).

[26] *CM* 67 f., *EP II* 169.

[27] *CM* 49 f.

[28] *Zur Phänemonologie des inneren Zeitbewußtseins*, *Husserliana* X.

[29] An account with more references is given in Th.M. Seebohm, "Reflexion and Totality in the Phenomenology of E. Husserl" in: *Journal of the British Society for Phenomenology*, Vol. 4, 1973.

objectifies itself. But the evidence in which it is given to itself, though apodictic, is not adequate in several respects. The first and most essential aspect of the structure which is "absent" is the actual emergence of inner temporality in form and content itself. It cannot be given in original evidence for objectifying acts in the *nunc stans*.

(3) Past phases of consciousness can be selectively thematized by acts of remembering. As past phases they belong to the immanence of consciousness, but they are for the acts of remembering a transcendence—a transcendence in immanence, as Husserl says. They are, furthermore, not given in Original and apodictic evidence in the same way in which consciousness is accessible to itself in reflections in the living present. Furthermore, the "past" is there and can even determine the activities of actual consciousness without being thematized by remembering. Indeed, only very few features are thematized and Husserl's analyses admit the recognition not only the unconscious but also of a subconscious realm within and for consciousness, i.e. a realm that is "absent" in principle and connected with the *nunc stans* which is as well "absent" for reflection.

The problem of the transcendence of the world can only be treated properly in connection with the analysis of intersubjectivity, i.e., the solution for the postponed problem of solipsism.[30] Something can be said about the transcendence of the world and objects in the world from the viewpoint of the transcendence in immanence mentioned above.[31] Transcendent objects are given as real objects of perception only via intentional acts of consciousness that synthesize disjunct phases of inner time in which the "same" object was given and may be given. Such intentional acts are the acts of remembering, expectation, and the presentification of the present absent—an act of the imagination that already includes the givenness of the other. It is, hence, the "transcendence in immanence," i.e. in the residuum of the reduction, which in part explicates the nature of the transcendence of identical objects of consciousness and their categorial structure—a point which cannot be pursued here.

Husserl's own account of the givenness of the other and intersubjectivity in the fifth Cartesian Meditation is difficult to understand because its background, the transcendental aesthetics of phenomenology, is nowhere explicated in the Meditations. It is clear, however, from the Meditations that Husserl recognizes that the transcendence of objects of perception presuppose that they are in principle possible objects for another consciousness, i.e. transcendence in general presupposes intersubjectivity. The consciousness of the other is, however, itself in a very peculiar sense transcendent for my consciousness. The analysis of the givenness of the other is, hence, a key problem for the "critique of the critique." We begin with a brief explication of the background that is, as mentioned, missing in the *Cartesian Meditations*.

---

[30] *EP II*, 173; *CM* 69, cf. the Addendum 239.

[31] *CM* 80; see also *Analysen zur Passiven Synthesis*, *Husserliana* XI, 112 f., (Henceforth *APS*), *KR* 189, *CM* 134, 140, 155.

The hyletic field, i.e. the "sensuous" contents of consciousness, belongs to the immanence of consciousness. The hyletic contents are not isolated data but rather given in a field that has its own constitution. This constitution is passive, and it is pregiven to all intentional activities of the ego. One aspect of this constitution of the field is the form of succession, i.e., the constitution of inner time consciousness. The other forms are the forms determining simultaneous coexistence in the living present, i.e., the field of localization.[32] The generally determining structures are Gestalt structures, i.e. contrast/fusion, pairing, foreground/background, etc. They determine, likewise in pure passivity, the associative awakening of past phases of consciousness.[33]

The ego has no active part in this constitution. It is rather affected by it. It is affected primarily by contrast-phenomena, i.e., not by the hyletic contents but by the relations they have with each other. The affection comes (1) from the field of localization in the living present, but also (2) from the past through associative awakening of contents by contents. The ego can respond to such affections with kinesthetic perceiving, remembering and expectations, activities which are closely connected with each other. The relation of affection between the hyletic field and the ego takes place within the field itself. The ego is in the field.[34] The ego's being in the field is determined by the most basic pairing relation through which the field is structured as a whole, namely the "inside" of my body and the "outside" of the field that surrounds it. The relation is not only a relation of localization. Temporal structures also play a role.[35]

An absolute here-now "inside" my body relates to the "there-outside" in the following way. A "there" is a potential "here" only in the retentional or protentional horizon of kinesthetic activities. That means that the here-now is always unique at every given moment, and it is at every moment confronted with a manifold of simultaneous theres which are potential heres, but never simultaneously actual heres. It should be noted that neither "here" nor "there" nor "here-now" can be understood as "points." They are all constituted by contrast-phenomena grounded in qualitative contents of the hyletic field. The "here-now" has as body a specific "thickness." The primordial ego in the field of its ownness, i.e., the ego that is left if one abstracts from the givenness of the other[36] is neither more nor less than the "pole," the center of the embodied herenow in the hyletic field. Its kinesthetic movements are the movements in which "theres" have been run through, made into here and let drop out the here-now again.

The other is, therefore, primarily experienced as a simultaneously given other here-now, i.e. another body. It is clear from the structure of the field and the nature of kinesthetic movements that such a simultaneous here-now can never be brought to original evidence, i.e., it can never be presented, it can only be appresented in my

[32] *APS* 133.

[33] *APS* 120, 205.

[34] *APS* 149 f.

[35] *CM* 137.

[36] *CM* 131–136.

original here-now. What is given is the outside of another living body. It is appresented as living because an inside is associated with it through an associated transfer of the inside/outside structure of my primordial body.[37] The associative transfer happens in the hyletic field in pure passivity, i.e., no activity of consciousness is implied. It "happens" to it, affects it. It experiences in passivity another "transcendence in immanence" that, in the dimension of localization, is the counterpart to the "transcendence in immanence" that confronts consciousness with the past.[38]

The givenness of the other as other living body in purely passive associative constitution is the ground for all higher forms of communicating with others and understanding them. Its first basic function is, however, a function in the representation of transcendent objects of perception. In the experience of a transcendent object of perception there are always only appresented, but not originally given, aspects of the object. To the extent to which they are appresented as simultaneously existing with the originally given aspect it is also appresented that they can be given simultaneously in original evidence for another consciousness in a here-now that is, seen from my point of view, there. Thus, transcendent objects are from the very beginning transcendent only as intersubjectively given objects. Their transcendence is constituted in the double transcendence in the immanence that is found on the one hand in the flowed off past of consciousness temporally and the transcendence in immanence in the field of localization with the givenness of the other. In both cases such constitutions are not the consequence of intentional acts, spontaneities of the subject. They "happen" to consciousness in passive constitution.

Some modifications of the concept of "apodicticity" and of the concept "transcendence" are necessary after these analyses, which belong to the "critique of the critique." First, one must distinguish between the apodicticity of the self-givenness of the existence of consciousness and the apodicticity in which certain structures of consciousness are given. In both cases apodicticity qua impossibility of not being given can be experienced only to the extent to which that which is experienced can be rediscovered in the structure of reflection on this experience. It is, hence, temporality that makes possible, such an experience because iterated reflection is possible only in the framework of temporality.

Apodicticity in the self-givenness of the existence of consciousness taken for itself is "blind." Without explicating the structures of consciousness, it does not lead to any content of critical knowledge. Having in mind the paradox of subjectivity and Heidegger's criticism, one has from the outset to avoid a serious error Husserl fell into and was, hence, wide open to the objection raised by Heidegger. The apodicticity of the self-givenness of consciousness to consciousness with respect to its existence ought not to be understood as an ontological statement. It is simply wrong to say that subjectivity is in any sense an "absolute" being or *"absoluter Seinsboden."* The statement about the apodicticity of self-givenness is a purely epistemic statement about a mode of knowing and not an ontological argument concerning existence. It is, therefore, also of interest only for a consciousness that has an interest in

---

[37] *CM* 137–145.

[38] *KR* 189; *CM* 134, 140, 155

apodictic evidence and such an interest is an epistemic interest. To ask for the "being" of the subject has as its immediate consequence that subject and object are not considered as correlates of their being known. This means, however, that talking about being is impossible inside the reduction. The reduction which is at least in the beginning the decision for one part of the paradox is cancelled if the concern is "being,"—the "being" of the subject is "being in the world."

The pure apodicticity of self-givenness with respect to existence taken for itself is useless seen from an epistemic point of view. According to Husserl there are apodictically given structures in an "eidos ego," "consciousness in general": the general structures of intentionality, temporality, actuality and potentiality, genetic structures and structures of the hyletic field. They serve in further investigations as parameters, frames for variations in which further structures of consciousness can be discovered. Such secondary structures are not given in apodicticity.

Structures of consciousness given in apodictic evidence are given as such only for a consciousness that has an interest in apodictic evidence. Furthermore, it holds that not all of them belong with eidetic necessity to all possible forms of consciousness. To the extent to which apodictic evidence is given with respect to intentional activities of consciousness such structures belong with eidetic necessity only to the type of consciousness that lives in such activities. Thus, all the apodictic evidence that can be discovered in a phenomenology of reason belongs with eidetic necessity only to a consciousness which has fully developed reason. Thus, there is only one realm in which "structure given in apodicticity" and "belongs with eidetic necessity to every type of consciousness" are not separable. It is the realm of passive constitution.

Regarding these structures, some further modifications in the concept of "apodicticity" must be introduced. They have immediate significance for the problem of transcendence. Prima facie it can be said that an apodictic evidence is, though not adequate, an original evidence. According to many interpreters and critics of Husserl who have enough textual evidence to support their claims, it could be said that according to Husserl an original evidence is an evidence given "bodily" in the living present and hence we have a metaphysics of presence. This interpretation does not fit however, the results of the analyses of passive synthesis. The result here is the apodictic evidence that there is for consciousness a necessary transcendence, absence in three dimensions: (a) the *nunc stans* as the place of its emergence is impenetrable, (b) the past of the passively constituted continuum of retentions and its contents as a transcendence in immanence and, (c) finally, the "there" in the field of localization as a simultaneous other "herenow" as another transcendence in immanence. What is "originally given" is absence as necessary absence. There should be no surprise about that. Those who find great delight in pointing to the priority of "absence" nowadays want to tell us as well that this priority is a necessary priority and cannot be cancelled, i.e., they claim to have apodictic evidence about this priority of absence.

Such an interpretation seems not to fit into the strict separation of the experience of the transcendent world from the experience of the transcendental •sphere of consciousness that Husserl introduces still in the beginning of the *Cartesian Meditations*. Furthermore, it does not fit any ontological understanding of the so-called phenomenological idealism. There can be no doubt that Husserl himself cherished idealism. There can be no doubt that Husserl himself cherished such a self-interpretation of his phenomenological investigations. But following his own analysis of the experience of the other and applying what he has said about the transcendence in immanence, the only possible conclusion—never drawn explicitly by Husserl and at some places denied—is that the difference of the "here-now" and the "there" in the hyletic field (and with it this field as a field of localization) belongs to the sphere of immanence, i.e. the transcendental sphere. That means that there is not only an "inner temporality" but also an "inner spatiality" left in the residuum of the *epoché*. It seems to me that without such an assumption it is impossible to give a phenomenological account of the givenness of the other as independent from the constitutive activity of my consciousness and hence of transcendence in general.

There will be, however, no difficulty for such an interpretation if the reduction is understood from the outset as only epistemic and not ontological. To "bracket" the natural attitude does not mean that the transcendence of the world has been bracketed. The reduction only "brackets" the claim that the world, in whatever understanding it presents itself, is the first evidence underlying all other evidence. The world, including its transcendence, remains as a validity phenomenon. The task is to analyze the meaning of this transcendence, beginning with consciousness as first evidence. If the reduction is understood as having only an epistemic function the following thesis is completely compatible with such a reduction: Transcendence emerges for the subject on the genetically lowest level of constitution in pure passivity. It emerges from the very beginning as something given for consciousness that is prior to all acts of consciousness. Seen under the reduction as an epistemic reduction the priority of the active life of consciousness is epistemic and not genetic. To state that consciousness can strive for evidence only under the genetic presupposition of a transcendence that makes such a striving possible, is itself an epistemic statement and as such even apodictically evident.

It can now be asked what phenomenology can say under these circumstances about the paradox of subjectivity. If the priority of the subject is understood only as an epistemic priority, then there will be no difficulty any more with the obvious observation that the natural attitude is genetically prior to the transcendental attitude. It is a general "law" of intentionality, that all intentions of oblique reflective nature are one-sidedly founded in intentions directed towards objects. Furthermore, the genetic priority of the natural attitude is—again in genetical investigations—shown to be a necessity. It is a necessity because it emerges out of the "transcendence in immanence" that determine the structure of subjectivity in passive constitution. It is the natural attitude that represents the other part of the paradox of subjectivity for which the being of subjectivity is a being in the world. The ontological priority of world and transcendence on which this part of the paradox insists is recognized under the reduction in epistemic interpretation. The priority occurs here as genetic priority, i.e., it is recognized that the "question of being" has indeed genetic priority in the genesis of knowing.

Finally, transcendental phenomenology can at least point to the original emergence of the whole context. The "*nunc stans*" was in its actual existence beyond the power and—metaphorically speaking—behind the "back" of phenomenological reflection. It is the common ground of all apprehension and self-apprehension and the source of the emergence of reality—as given with or without the attitude generated by the reduction. Seen from an ontological point of view the urge to say something about it is pressing. Seen from an epistemic understanding of the reduction one has simply to state: in this case the presence is given throughout in absence.

I tried to give a version of transcendental phenomenology under which the objections of Heidegger (and others) in the *Grundprobleme* and later are rendered pointless. There is, however, in the *Grundprobleme* a positive claim which implied another criticism. It is impossible, according to Husserl and his reflections, to achieve ultimate grounding in the natural attitude, i.e., under the part of the paradox that explicates the being of subjectivity as being in the world. Heidegger apparently thought otherwise, at least in the beginning of his career. Let me point out in the last part of this paper that Heidegger's *Denkweg* leads, in its end, precisely to the result which can be foreseen from Husserl's point of view. This is, of course, not something that has been said by Heidegger in the way in which I state it here. Rather it is part of his later universal criticism of transcendental philosophy.

Kockelmans has shown[39] that Heidegger's intention in *Being and Time* and other writings belonging to the same period is still to present an epistemology which has the task of a philosophical grounding of the sciences. Fay[40] has collected material that shows that at this time Heidegger had a similar attitude with respect to formalized logic and mathematics. There is also no hint at this time that the idea of ultimate grounding via the explication of the ontological difference has been abandoned. Careful reading with hindsight, i.e., having in mind the later writings of Heidegger, allows us, however, to determine the problem which later forced Heidegger to rethink the problem of grounding and ultimate grounding.

We find in *Being and Time* first a "more profound grounding" in the explication of temporality as the "ontological meaning of care." In this explication "theoretical knowing" occurs in the transition from the second to the third level of temporality.

The first level of grounding is the analysis of the temporality of the being of Dasein. Here "resoluteness comes back to itself in the future and posits itself in the present of the situation."[41] "Time is originally the timing of temporality and as such the possibility of the constitution of care."[42]

The second level is the level of the "*abkünftige*" temporality. The word "*abkünftig*" is not an original German word. It means on the one hand "grounded," "secondary," but it has, on the other hand, the connotation of "turning away from the origin" which is, as mentioned, future, "*Zukunft*" as "*Abkunft*." On this level of time, temporal beings are given. It is the time of circumspect care in the structure of being-concerned-with. As ecstatic time which has no serial character it has the

[39] Joseph J. Kockelmans, *Heidegger and Science* (op.cit.).

[40] Thomas A. Fay, *Heidegger: The Critique of Logic*, The Hague, Nijhoff 1977.

[41] Heidegger, M., *Sein und Zeit,* Max Niemeyer, Tübingen, 1953, 326 (Heceforth *SZ*). The translations are mine.

[42] *SZ* 331.

modes of "now as," and "then, when" in the past and in the future. Time is public time on that level as the work of the days and nights disclosed in circumspect care.[43]

The third level is the level of abstract world time. It can be characterized with Aristotle: "For time is this: what is counted in the movement in accordance with what is earlier and what is later." Time on this level is a counted series of nows and instead of "places" one has the coexistent points of abstract space. Both are indifferent to being concerned with," and, hence, meaning.[44]

Heidegger introduced the existential conception of science in the analysis of "circumspect care." This existential concept of science is different from the concept of science that understands science as *Begründungszusammenhang*, as connected grounding. Science is understood as a mode of being in the world, in which beings (*Seiendes*) and Being (*Sein*) are discovered or disclosed. The existential conception of science is explicated by asking for the ontological genesis of theoretical behavior and the theoretical history of science.

The relevant point is not the simple vanishing of the practical "being concerned with," i.e., the theoretical attitude is not reached by simply no longer being involved in things with which we are "being concerned with." The origin of the theoretical attitude is rather a "looking around" that has the character of "checking what is the case" (*Nachsehen*). It occurs in the "being concerned with" and thus the "being concerned with" has its own "theoretical aspect." This explains why later theories always have a practical aspect in their techniques of observation and experimentation.

What happens in the transition of the second to the third level of temporality is the discovery of "if-then" relations in the "being concerned with" and "circumspect care" of the second level. This relation receives the character as a relation between facts (*Vorhandenem*) and with this objective grounding relation we have science understood as "*Begründungszusammenhang.*"

This brief reminder is sufficient to illustrate two points. (1) Though the style is different, the whole analysis reminds us of Husserl's theory of the levels of time constitution. One could, indeed, try to understand the whole analysis in a phenomenological framework, add some corrections from the phenomenological point of view, but also enrich the phenomenological analyses given by Husserl with viewpoints taken from Heidegger. An epistemic reflection on the analysis itself would then immediately lead to the reduction. (2) This will be, however, a gross misreading of Heidegger's intentions. The whole analysis of Dasein in *Being and Time* has only the function to prepare the question of being which discloses the ontological difference. According to the *Grundprobleme* the ultimate grounding can take place only there.

In *Der Satz vom Grund*[45] Heidegger finally abandons the self-interpretation of the ontological difference as "ultimate grounding" in the traditional sense in his criticism of the principle of sufficient reason. Already in *Being and Time* Heidegger

---

[43] *SZ* 406 f.

[44] *SZ* 421.

[45] *Der Satz vom Grund*, Neske, Pfullingen, 1957, (Henceforth *SG*.)

points out that the if-then of circumspect care becomes a universal principle if it is applied to pure objects (*Vorhandenes*). The environment is thus changed into nature *qua* nature of the natural science. In its original form the "principle of sufficient reason" simply states that "everything has its reason," and "Nothing is without a reason." It has its time of "incubation" in the time of metaphysical thinking. In this metaphysical thinking it has, according to Heidegger, a "double tune (*doppelte Tonart*)." Let us simply say it has a double meaning.

The first meaning is expressed by the "*principium reddendae rationis sufficientis*," the principle of sufficient reason. In this form it leads metaphysics to the "causa prima" which is "deus."[46] The second, more hidden meaning discloses itself in the "speaking of being which grounds the beings." At the end of the time of the incubation, the first meaning is the governing meaning that has pushed the second completely into the background. In this governing meaning the principle of reason governs science, mathematics and technology. Reason is understood here as "reckoning ratio" and logic as the calculus of mathematical logic. The goal is universal rationalization, i.e., the delivery of reasons we can reckon with, count on.[47]

The striving of modern philosophy for "ultimate grounding" belongs, according to Heidegger, to the unfolding of the first meaning of the principle. Pure reason is in Kant as theoretical and as practical reason *ratio pura*, ground of all grounding and positing of the absolute ground.[48] Kant's method brings the principle of sufficient reason to its ultimate power. "It reaches its ultimate power in the principle of subjectivity which is nothing else than to claim unlimited power for the principle and its striving for complete rationalization."[49]

Heidegger criticized Husserl in the *Grundprobleme* for not asking the question of being. Without asking this question one cannot reach ultimate grounding. There is a shift in the *Satz vom Grund* which affected this criticism. To ask for "ultimate grounding" under the first meaning of the principle and thus also turning to the principle of subjectivity is itself the basic mistake. Heidegger does not mention Husserl in the *Satz vom Grund* but it is obvious that the characterization of Kant there is also valid for Husserl. If the "ultimate ground" is thought as subjectivity, then the "*Entzug,*" the vanishing of being and the vanishing of the possibility of speaking being, is the necessary consequence. To ask for "ultimate grounding" means that the question of being cannot be asked properly.

This verdict can be considered from two sides. It says as well that asking the question of being, asking the question of the being of the subject and answering this question with "being in the world" does indeed also exclude the possibility of asking meaningfully for ultimate grounding. This is precisely why Husserl bracketed the natural attitude and why I argued in this essay that already an ontological interpretation of the reduction destroys the possibility of reaching the goal that Husserl declared to be this goal: to find a meaningful explication of "ultimate grounding."

---

[46] *SG* 53 f.

[47] *SG* 53 f., 98 f., 166 f.

[48] *SG* 127.

[49] *SG* 138.

Is it possible that there is an ideal of "ultimate grounding" that corresponds to the second meaning of the principle of reason? First, it has to be said, according to Heidegger, that we still live in the time of incubation in which the "second tune" cannot be heard.[50] Secondly, it is obvious from the hints that Heidegger gives, that the speaking of being, die Sage des Seins, has the character of poetical-prophetical discourse in which a mystical experience of the ground as an abyss, the *Grund* as *Abgrund*, speaks. It is difficult, but not impossible, to translate the discourse of ultimate grounding in approximately one-to-one corresponding terms from one language into another. Thus, it is an indicator that Heidegger is after something which is still unknown and his hints in this respect in the *Satz vom Grund* cannot be translated. They can only be explained with the aid of philological commentary on some peculiarities of German.[51] Thus, Heidegger had to abandon the original program of the *Grundprobleme* that was connected with his criticism of Husserl, namely, to fulfill Husserl's program of ultimate grounding in asking first the question of the being of the subject and then the question of being.

Thomas M. Seebohm was professor emeritus at the Johannes-Gutenberg Universität Mainz. For a more detailed description of his career and his work, please consult the "Introduction" to this volume.

---

[50] *SG* 90, 97.

[51] *SG* 96, 98.—In German "principle" is "*Grundsatz*" or—as in Heidegger—briefly "*Satz*". "***Satz***", however, has a second meaning in German, which is disconnected from the first, i.e., "statement", namely "leap". "Abyss" is in German expressed by using the linguistic root "*Grund*" and called "*Abgrund.*" Thus, the German counterpart to "principle of reason" is "statement of the ground," which is rendered by Heidegger in the "second tune" to "*Satz in den Grund*" and then "*Satz in der Abgrund*", "leap into the abyss." The impossibility of translating this "movement of thought" has no longer anything in common with the difficulties of translating technical philosophical terminology.

Chapter 15
Possible "Worlds": Remarks About a Contoversy

Thomas M. Seebohm

Abstract J.N. Mohanty is one of the few phenomenologists who has been able to enter a fruitful discussion with analytical philosophy. Most significant in this respect are his writings on Frege. But there is another area which also deserves attention. It is his discussion of Jaakko Hintikka's attempt to explicate Husserl's concept of intentionality in terms of intension, more precisely, the concept of intension which has been developed in the framework of the so-called possible world semantics. The outcome of his analysis taken from a phenomenological semantics on the level of predication to problems connected with the "possible worlds semantics" was that the "world" as it is to be understood in the context of formalized logic cannot be understood as some kind of concrete world, the real world in which we live, the "worlds" in quasi-positionality or worlds constructed in metaphysical speculations as correlates of other intellects. They are the result of a peculiar abstraction and idealization which separates them in principle from all other conceptions of world. Thus, the paradoxes that result from the attempts to connect the different concepts of "world" vanish. Mohanty also show how another problem occurs which is a result of the separation. How can the explication of modal operators and other operators – as well as some intensional relations by means of this abstract apparatus – have objective validity for the explication of some categorical forms used in ordinary talk about the real world in which we live? Here we have a question about "objective validity" which belongs to transcendental logic. In the context of Husserl's phenomenology that means: it must be solved by means of an analysis of the genesis of these categorical forms in the pre-predicative realm.

This chapter was previously published in the following book:
Possible "Worlds," in Phenomenology East and West, ed. FM Kirkland & DP Chattopadhyaya, pp. 129–148 (Springer, 1993)

T. M. Seebohm (deceased)
Mainz, Germany

© Springer Nature Switzerland AG 2020
T. Nenon (ed.), *Thomas Seebohm on the Foundations of the Sciences*,
Contributions to Phenomenology 105,
https://doi.org/10.1007/978-3-030-23661-8_15

This chapter reviews J.N. Mohanty's analysis of Jaakko Hintikka's attempt to reformulate Husserl's notion of intentionality as intensionality conceived along the lines of analytical philosophy's discussion of possible worlds theory. It shows that the outcome of applying viewpoints taken from a phenomenological semantics on the level of predication to problems connected with the "possible worlds semantics" as considered in the analytical tradition was that the "world" as it is to be understood in the context of formalized logic cannot be understood as some kind of concrete world, the real world in which we live, the "worlds" in quasi -positionality or worlds constructed in metaphysical speculations as correlates of other intellects. Phenomenological reflection shows how they are the result of a peculiar abstraction and idealization which separates them in principle from all other conceptions of world. Thus, the paradoxes which are connected with the attempts to connect the different concepts of "world" vanish when reflections on the genetic origins of these kinds of abstractions are undertaken.

J.N. Mohanty is one of the few phenomenologists who has been able to enter a fruitful discussion with analytical philosophy. Most significant in this respect are his writings on Frege. But there is another area which also deserves attention. It is his discussion of Jaakko Hintikka's attempt to explicate Husserl's concept of intentionality in terms of intension, more precisely, the concept of intension which has been developed in the framework of the so-called possible world semantics. Fourteen years after Mohanty's famous paper[1] the situation has changed. Thus it is meaningful to reconsider some aspects of the controversy. Before doing so a brief recapitulation of some points of the controversy is necessary.

Hintikka's significant contribution in the *early* phase of the development of modal logic was to transform Quine's well-known criticism. Quine pointed out that the operators introduced by modal logic had an affinity to particles expressing propositional attitudes, i.e., particles followed by "...that ('p')", and furthermore violated the principle of strict extensionality by assuming that we can speak about essences and intensions in the language of modal logic. Searching for "models for modality", Hintikka used just these two points in order to show the universal applicability of the new post classical logic. What was considered to be a weakness was now used to show the strength of the new formalism. His attempt to include also Husserl's concept of intentionality is, as Mohanty pointed out, a specific extension.[2] All other applications are applications – I use phenomenological argot here – on the predicative level, i.e., on the proper level for logical investigations. Hintikka argued, however, that also the intentionality on the level of prepredicative perception is a model for modalities.

Mohanty admits the possibility of an application of the technical apparatus of possible world semantics for two cases which can loosely, and with qualifications,

[1] The paper read at the meeting of the American Philosophical Association, Eastern Division, Washington, D.C. 1978 is a critical discussion of Hintikka, J. 1969. *The Intentions of Intentionality and Other New Models for Modalities.* Dordrecht: Reidel. It was published with the title "Intentionality and Possible Worlds: Husserl and Hintikka" in *Husserl, Intentionality and Cognitive Science,* ed. H.L. Dreyfus, (Cambridge: MIT Press, 1982), p. 233–55.

[2] Mohanty, *op. cit.,* p. 238.

be connected to the phenomenological concept of intentionality. Fregeau senses can be understood as "counterparts" of Husserlian noemata on the predicative level, i.e., they are noemata connected with a sign and thus the sense of the sign. Their directedness towards the object can then be reconstructed *also* in terms of possible worlds semantics, though this reconstruction *cannot* count as an explication of Husserlian intentionality.

> To know the meaning of a term – say "red" – is to know what it is to be red, which includes the ability to recognize a thing as red under various different circumstances. That the concept of the "meaning" of a term has something to do with that of "possibility" – and both with that of "essence" – cannot be denied.[3]

Mohanty also admits that descriptions of the form "I perceive ('p')" can be treated meaningfully in terms of possible worlds semantics. Given such a description all possible states of affairs can be divided in two groups: those that are compatible with p and those that are not.[4] According to Mohanty, this statement is true and also forms the main support of Hinttikka's appeal to "possible worlds". But it does not describe perceptual experience. Rather it concerns what must be the case if a perceptual statement is true. The statement tells us something about the truth conditions of statements about perceptions at the level of formal logic. It is, however, not a phenomenological statement and it contributes nothing to the phenomenology of perception. Mohanty adamantly denies that the concept of "possibility" – which in the analysis of intentionality occurs as the indeterminacy of the horizon, especially at the level of perception – can be explicated in terms of the "possibility" of possible world semantics. Given the scope of the discussion he could not give an explicit account of the nature of this "possibility", i.e., an account of the phenomenology of "modalising".[5] A careful investigation which takes as its point of departure the *Analysen zur passiven Synthesis* would prove his point. Since such investigations also have the goal of clarifying genetically the origin of categories – including modal categories – on the prepredicative level, some results concerning the ontological difficulties of individuals, identity and essences in possible worlds can be discussed under these viewpoints. The task in this case would not be to point out the shortcomings of an explication of the phenomenological theory of perception as a "model of modality" but rather to offer a phenomenological critique of modal categories, including those treated in modem modal logic.[6]

---

[3] *Ibid.*, p. 235.

[4] *Ibid.*, p. 241.

[5] Edmund Husserl, *Analysen zur passiven Synthesis,* Husserliana XI, (den Haag: Martinus Nijhoff, 1966), Erster Abschnitt, "Modalisierung", pp. 25–64; cf. Edmund Husserl, *Erfahrung und Urteil,* (Claassen und Goverts, 1948), III. Kapitel, "Der Ursprung der Modalititen des Urteils", pp. 325–80; *Experience and Judgement,* trans. J. Churchill and K. Ameriks, (Evanston: Northwestern University Press, 1973), Part II, chap. 3, "The Origin of the Modalities of Judgement", pp. 271–313.

[6] My considerations concerning this problem will be published in *Phtinomenologische Forschungen* and the *Journal of the British Society for Phenomenology.*

The topics I want to consider in this essay are Mohanty's abovementioned qualifications. (1) Mohanty does not admit that anything is gained by applying the technique of possible worlds semantics to phenomenological investigations or by explicating phenomenological concepts with them. What he admits is that, seen from a phenomenological point of view, nothing is wrong in the application of this technique for special cases in logical investigations. (2) Nevertheless Mohanty has strong reservations about the "ontology" connected with this semantics. He uses a quote from Hintikka himself to illuminate the reasons for his doubts.

> It would be more natural to speak of different possibilities concerning our 'actual' world than to speak of several possible worlds. For the purpose of logical and semantical analysis, the second locution is more appropriate than the first, however, although I admit that it sounds somewhat weird and perhaps also suggests that we are dealing with something much more unfamiliar and unrealistic than we are actually doing.[7]

In his response, Hintikka seems to assume that Mohanty's ontological doubts are connected with the reification of possible worlds, and denies that he defends such a reification.[8] In other words, Hintikka rejects the "realism" of D.K. Lewis.[9] However this cannot be all that Mohanty has in mind. In section III of his paper, we find an extended analysis of Husserl's theory of quasi-positional acts and the "fictional entities" which are the objects of such acts. According to the conceptualist interpretation of possible world semantics, other possible worlds are represented in fictions. Mohanty claims that Husserl' s quasi-positional worlds lack unity of contexts and a unified framework, and for this reason it must be said that identical individuals in the proper sense exist only in the actual world. Thus Husserl's quasi-positional worlds – though they are constitutive for a certain conception of "possibility" – cannot be explicated with the "possible worlds" of the new semantics.[10] In his response, Hintikka claims further for his interpretation of "possible worlds" that "we cannot even hope to operate actually with any possible world as a completed whole."[11] Seen from a certain point of view this is true, especially for his own specific "linguistic" interpretation. If a possible world is represented in a maximally consistent set of statements, then it can be assumed that this formal idea of such a completed set may be represented with fictional content only by a small subset of statements. All others are added only via the formal rules of construction for such sets which do not require any knowledge of the actual fictional meaning of the statements added. Thus, seen from this viewpoint, he can indeed claim that a possible world is not fully determined in his interpretation. The main problem is not that Hintikka's interpretation can nowadays be considered too narrow, because it excludes some

---

[7] Mohanty, *op. cit.*, p. 235.

[8] Hintikka in his response in *Husserl, Intentionality and Cognitive Science, op. cit.*, p. 253.

[9] Lewis, D.K. 1973. *Counterfactuals.* Cambridge: Cambridge University Press. Here Lewis defends "realism". For Saul Kripke and others "conceptualism" is more convincing. For a survey of positions and references, see my *Philosophie der Logik.* (Freiburg: Karl Alber Verlag, 1984), pp. 202–218.

[10] Mohanty, *op. cit.*, p. 245.

[11] Hintikka, *op. cit.*, p. 254.

technically interesting developments.[12] Rather, seen from the viewpoint of a phenomenological "semantics", the basic problem is its ambiguity. Though we do not know what the statements tell us, nevertheless, according to the rule of construction, we have to assume that they are "true" with respect to a world and hence also that this world is fully determined. It is fully determined in the same way in which our world, considered as a correlate of a maximally consistent set of statements, is considered to be fully determined in classical semantics. That means, however, that identical individuals are formalistically supposed to exist in this possible world as well as in our world. This ambiguity can be considered today under some viewpoints which were not yet given ten years ago.

There are at least some logicians and philosophers belonging to the analytical tradition for whom some of Quine's doubts concerning modal logic are not removed but rather confirmed in the development of the semantics of possible worlds, especially on the level of predicate logic. Already the controversy which surfaces in Mohanty's essay[13] between Platinga and Hintikka about transworld or worldbound identity of individuals indicates that it might be "hopelessly obscure what it was that quantified modal logics quantify over."[14] The "referential opaqueness" of modal propositional logic leads to a metaphysical jungle, an ontological slum.[15] For Quine the jungle consists in the very possibility of admitting some entities about which the language of classical logic could not speak at all, e.g., essences. However, recent handbooks show where the real problem is. They mention many classes of models for predicate logic, many ways of distributing individuals to worlds and of defining stronger or weaker types of identity for different systems of predicate logic under one class of models, e.g., SS. The problem is that the formalism as such allows us to construct formally correct semantical systems which imply different ontological assumptions. For example, formal systems can be constructed in a way which admits individual essences but also in a way which does not. They can be constructed with worldbound individuals, with individuals having transworld identity, and also with the possibility or not of talking about essences, and all of this can be done with many modifications. The language of classical logic has an austere but rigid and unambiguous ontology if it is understood under the objective interpretation of the quantifiers. However, the technical apparatus of the semantics of modal logic provides us with several kinds of options, and thus the formalism does not *per se* answer ontological questions. Such questions can again be topics for traditional philosophical disputes. In other words, instead of logic as philosophy we have a philosophy of logics.[16]

---

[12] D.K. Lewis, *op. cit.;* cf. Montague, R. 1974. *Formal Philosophy,* ed. R. Thomason. New Haven: Yale University Press, 148ff.

[13] Mohanty, *op. cit.,* p. 242.

[14] Haack, S. 1978. *Philosophy of Logics.* Cambridge: Cambridge University Press, 194.

[15] W.V.O. Quine, "Three Grades of Modal Involvement", in *Proceedings of the XI International Congress of Philosophy* (14), p. 174. Cf. S. Haack, "Lewis's Ontological Slum" in *Review of Metaphysics XXX* (1977), pp. 415–429.

[16] Cf. S. Haack, *Philosophy of Logics, op. cit.*

A further problem is that these options are "suggestive" for specific treatments of ontological problems connected with the status of possible worlds. It is already obvious from the discussion between Mohanty, Hintikka and – in the background – Platinga, that a transworld identity of individuals is not "suggestive" of a reification of possible worlds, but a worldbound interpretation is very suggestive for such an interpretation, as the example of D.K. Lewis shows. Thus Hintikka's defense of ontological ideas, which are somehow close to Mohanty's and Husserl's in his eyes, is not a matter of possible worlds semantics as such. He has rather chosen a function of assigning individuals to worlds which admit certain assumptions concerning entities and would have been excluded by choosing another assignment function.

What happens here on the level of predicate logic is foreshadowed on the level of propositional logic. "Accessibility relations" between possible worlds seem to be innocent abstract mathematical entities. But the different propositional attitudes are defined precisely by means of the accessibility relations. There are now some "propositional attitudes" which immediately demand that "possible worlds" have to be understood as index points in our world, and the accessibility relations connecting them are relations in the actual world, e.g., temporal relations.[17] There are others, like the alethic modalities which require or are more "akin" to the so-called possible world realism of D.K. Lewis. Thus the problem of the ontological ambiguity of the formal machinery of the possible worlds semantics surfaces already on the propositional level.

Husserl's phenomenological semantics, the "logic of truth" which he also calls the theory of modalising,[18] has a prepredicative and a predicative level. The prepredicative level of passive synthesis and the synthesis of imagination have a structure which determines the categorical structure of the predicative level. The analysis of this level is "semantical", because the protocategorical forms have their genesis in structures of positive and negative fullfillment. The determination of the categorical structures of the predicative level by the structure of the prepredicative level "explains" the objective validity of the categories, i.e., it shows that such a Kantian problem of the objective validity of categories "solves itself" in the context of a Husserlian transcendental aesthetics which is at the same time the foundation for a transcendental logic. It is, however, sufficient for the limited purposes of a phenomenological analysis of certain difficulties in formal logical semantics to restrict the considerations to the predicative level. The reader has only to keep in mind that the concept of the "real world in which we live" presents a reality which is in the final instance grounded in passive synthesis in its content and form and is, hence, given independent of all "acts" of the subject.

Unqualified "truth" on the predicative level is the complete correspondence *(Deckung)* of the state of affairs meant in the judgment and the state of affairs, if and only if the judgment is given in perfect predicate evidence, i.e., in perfect categori-

---

[17] See for example Prior A.N. 1967. *Past, Present, and Future.* Oxford: Oxford University Press.

[18] Cf. Edmund Husserl, *Analysen zur passiven Synthesis, op. cit.,* Abschnitt II, pp. 65–100 and *Formal and Transcendental Logic,* trans. D. Cairns, (The Hague: Martinus Nijhoff, 1969), chapter 5.

cal articulation, and the state of affairs is given in perfect prepredicate evidence. Unqualified falsehood is given under the same conditions if there is a discrepancy between the state of affairs meant and the state of affairs. It is often not seen that "evidence" in Husserl refers not only to the present and "truth" but in the same degree also to the absent and "falsity". In the judgmental activity of the lifeworld, but also of the sciences, truth and falsity are "modalised". That is, even if the predication is affirmative and no modal operator occurs in it, the predication is in the overwhelming majority of the cases accompanied by some insecurity which is not articulated categorically. The same holds for negative judgments as well. A more thorough analysis of this fact would have to tum to the prepredicative level and the role which expectation, confirmed expectation, disappointed expectation, disjunctively split expectation, etc. , play in the genesis of logical forms.

What has been said belongs to the phenomenology of doxic-thetic judgmental activity in the lifeworld, i.e., to the judgmental activity in "ordinary language". Taken for itself it has no significance for formal logic, with the exception, perhaps, that it indicates clearly that our experience of the real world in which we live is in every respect "modalized" and that this "modalization" has many different forms which are expressed on the predicative level with particles indicating "propositional attitudes". We reach the level of classical logic through (1) a reflection on the difference of state of affairs meant and state of affairs in general, (2) a formalising abstraction, and (3) an idealisation. (2) and (3) are onesidedly founded in (1). The formalising abstraction "grasps" the categorical form and represents all the contents which are given in this form – the *kategorische Stoffe* belonging to this form, e.g. , sentences in the case of sentence connectives – to be represented by "variables". The idealisation considers for all substitution instances of the variables only unqualified truth and falsity in the sense mentioned, i.e., it eliminates "modalization". Under these presuppositions the two logical principles of non-contradiction and the excluded middle are valid without qualification: "a statement is either true or false, never both, and a third is not given". Traditional non-formalised logic is grounded in the same type of idealisation. What is missing is a completed formalising abstraction. Thus traditional logic is not able to spell out the necessary consequences, i.e., the definitions of the connectives representing categorical form by means of the principles.

Seen from the point of view of phenomenological semantics the basic question for the idea of a semantics of possible worlds is: what has happened to the concept of world under the idealisation which is the basis of classical logic? Since the semantics of possible worlds is nothing but an extension of classical logic, its concept of world will depend on the concept of world which has been developed already on this level. Given this idealisation one can "think" the "world" as the sum total of the state of affairs which correspond to the set of true statements. Since the two principles oflogic hold, (a) there is no modal operator possible with respect to this world, (b) the set of true statements must be a consistent set, and (c) those statements to which nothing that is the case in the world corresponds are false.

This "world" of the logician, the Wittgensteinian sum total of everything which is the case, has the character of a world in itself, i.e., the world as it is without our subjective and relative thoughts and our always modalised judgments. Thus – and this is essential – this conception of "world" must be distinguished not only from the quasi-positional fictional worlds but also from the real world in which we live. *De re* modalities are characteristic for the ontological structure of the real world. It is, e.g., an open question whether there will be a sea battle tomorrow. The real world knows temporal structures. Quine recognized clearly that the "world" about which the language of formalized logic speaks does not admit propositional attitudes and hence also none of the temporal structures of the "real" world. According to Quine, the only candidate for a representation of the world of logic is the atemporal world of modem physics in which the individuals are the "points" in the four-dimensional spatio-temporal field and the predicates the events given in them.[19] It is essential to remember that the worldview of science – and especially physics – confronts us with worlds which are also "worlds in themselves" and different from the world as it appears to us, i.e. the real world in which we live. But not all of these worlds are candidates for being the world about which a logic, like classical logic with no knowledge of modal operators, speaks. It is, hence, wrong to identify the "world in itself" of logic with the "world in itself" of physics. The abstractions and idealisations which are essential for many physical theories are much less radical than the formalising abstraction and the idealisation which leads to the unqualified true/false dichotomy of classical formalised logic.

The "semantics of possible worlds" presupposes the basic abstractions and idealisations of classical logic. In the overwhelming number of classes of models which have been developed, the logical principles are valid without qualification with respect to each possible world. For every world, it holds that it can be understood as the sum total of states of affairs for which a certain set of statements is true. It is also assumed that this set will be a consistent set. But so called non-normal systems in which basic principles of logic are violated have been introduced.[20] There are other systems which have – for valid intuitive reasons as we have seen – truth value gaps. Such radical extensions, however, only show once again the ontological ambiguity of the technical apparatus.

Setting aside such extensions (which require abstractions going even further than those which determine the "world in itself" of logic), it can be said that the possible "worlds" of the new semantics are "logical worlds" in the sense which is predetermined in the concept of world of classical logic. This assertion is not grounded in any linguistic, ontological or epistemological interpretation of "possible worlds". It is simply a question of the techniques used already at the level of propositional logic. Seen from this viewpoint all other interpretations are always only in some aspects true and wrong in all other aspects. Their being "wrong" is in the final instance grounded in the equivocations in the concept of "world".

[19] Quine, W.V.O. 1977. *The Way of Paradox and Other Essays.* Cambridge: Harvard University Press, 228–254, especially p. 237.

[20] The difference between "normal" and "non-normal" was introduced by Saul Kripke. 1965. Semantical Analysis of Modal Logic II. In *The Theory of Models,* ed. J .W. Addison et al. Amsterdam: North Holland, 206–220.

The equivocation is given to the extent that the term "world" can refer on the one hand to the world which is the world in itself of logic and on the other hand to the world which is the real world in which we live. "Realism" with respect to possible worlds is tempting, because to assume that a possible world makes a certain set of statements true and its complement false is to "think" the possible world is a "world in itself" in the same way as the actual world of classical logic. It is a deeply rooted prejudice, which is again grounded in an equivocation, that the world "in itself" is the real world and the world in which we exist is an "appearance". This is to a certain degree correct with respect to the "world in itself" of science. It is meaningful to speak about the physical reality. But the "world in itself" of logic is already for classical logic the result of a much more radical abstraction. Hence the "world in itself" of logic no longer has anything in common with the world on which it is true – with the admixture of a low degree of uncertainty – that, e.g., the genetical make up of kangaroos is such that they have tails and do not topple over.[21] Thus the concept of world in the logical sense cannot be transferred to or be said to underlie the conception of the real world in which we live where there are concrete cases of statements that are materially true.

It is a mistake as well to assume that "possible worlds" are "real" in the sense in which logical and mathematical objects are supposed to exist in so-called Realism or Platonism. This can be accepted only as long as we consider the whole semantics of possible worlds as an ideal entity belonging to the types of ideal entities which we find in set theory. It is still in a certain sense possible to consider the function, which assigns sentences to worlds in which these sentences are true, not simply as a function but as a function which assigns truth values. However, in the framework of the so-called "realism" with respect to ideal objects, it is impossible to assume that this function assigns a concrete statement about some non-stubnosed Socrates to a fact in some possible world in which it is true. A possible world to which concrete statements are assigned as true in this world cannot be understood as an ideallogico-mathematical entity: it will be real not in the ideal sense but in a very real material sense. Secondly, as already mentioned, for the purposes of logic it is neither possible nor necessary to make any assumption about the actual truth value of some concrete statement in the "world in itself" of formal logic. It is enough for this purpose to assume that the principles hold for this statement.

With respect to conceptualism, Mohanty, as discussed above already, mentioned the basic difficulties connected with the interpretation of the possible worlds of logic as quasi-positional worlds of fiction. We discussed in the last paragraphs the difference between the real world in which we live and the "world in itself" of logic. But it is obvious that the quasi-positional worlds of fiction are quasi-positional "real" worlds. Husserl said that all doxic-thetic modalities which can be found in the real world are given in such quasi-positional worlds as well. Hence they cannot represent the "world in itself" of logic. It is very difficult to imagine what kind of mind could have possible worlds as its objects. The assumption of more than one

[21] This is a variation of the example which has been repeatedly used by O.K. Lewis.

world is rare in the history of philosophy. Setting aside the idea of a sensible and a supersensible world – which is for obvious reasons also not a candidate – it is Leibniz to whom those who defend a conceptualistic interpretation refer. It is assumed that one has simply to drop the metaphysical "dressing" of Leibniz's theory in order to reach a meaningful explication. One has to keep in mind, however, that it is this metaphysical viewpoint which makes the whole idea acceptable. Possible worlds are thought to be the object of the infinite intellect of God before the creation. In order to select the best we have to assume that the infinite intellect "knows" precisely what is materially true and what is materially false in each of them. How is this possible? The only acceptable answer is that God can have such a knowledge because he has an infinite intellect. But here the ball stops. We as finite intellects by definition do not know how an infinite intellect works. Thus we do not know how to know possible worlds which are different from our actual world in the way God knows them. God's knowledge of possible worlds is not grounded in either the abstractions which lead to the concept of logic's "world in itself" or the multiplication of this world into the set of "possible worlds". Rather he really knows them in the same way in which the divine providence materially knows everything which was, is, and will be in our actual world. There is no way from God's possible world to logical "worlds in themselves" and no way from them to quasi-positional worlds of fiction and vice versa. In order to start the construction of maximally consistent sets, no fiction is necessary. What is needed on the one hand is the assumption of a set of statements which are consistent. No other specifications are necessary. On the other hand, the further determination of a fiction is not interested in all statements which are consistent with the set of statements representing the fiction. The fact that certain minerals have been found on the moon is certainly consistent with the "fiction" represented by *Death in Venice,* but adds nothing which could be useful for the closing of gaps of information in it.

Every interpretation of "possible worlds" is absurd and bizarre if the attempt is made to understand it as a real or conceptual- fictional variant of the real world in which we live. Already the "world in itself" of classical logic has nothing in common with the real world in which we live. The "world in itself" of classical logic is nothing more than a set of states of affairs which can be described in a maximally consistent set of statements that are supposed to be true without qualification with respect to this set of states of affairs. Nothing more can be said about this world. Given this point of departure there is no difficulty in inventing other "worlds" of this sort. The assumptions that the complement of such a set has as subsets other consistent sets of statements and that such statements can be members of different sets is *afortiori* legitimate for possible worlds of logic, because they are understood first of all as sets of states of affairs. In general, everything which can be meaningfully assumed about sets can be assumed about possible worlds. It is, hence, not even necessary to insist that they are represented by consistent sets of statements.

Classical formalized logic and its ontology require that all states of affairs be describable only in terms of abstract individuals, i.e., individuals which have only one "property", namely identity, and which are the bearers of *n-adic* predicates. It is questionable whether individuals of this sort are given in the real world in which we live. An investigation in the prepredicative realm could show that this is certainly not the case. The problem of "names" or individual constants illustrates this fact on the predicative level. If "names" are admitted into the language of formalised logic, they have to obey requirements which names in natural languages never fulfill. The only way of fulfilling these requirements in natural language would be to "name" things with numbers in order to check that no "name" refers to more than one individual. Names in natural language refer to one thing only if they are context bound. That means, however, that they refer to individuals via n-adic predicates. The opinion of some philosophers[22] who say that the individual *is* the sum total of its real predicates doubtless has its own difficulties. It is, however, an opinion which is grounded in structures which belong to the real world in which we live.

Quine's interpretation of the individual variables on the level of ordinary language, as well as his presentation of the world of logic as the world of ideal science, illustrates the abstraction through which we get individual variables well. What corresponds to them in ordinary language are demonstrative pronouns, i.e., indicators of an empty place in which something is given. The individuals of his four-dimensional world are space-time points. Formalised classical logic, however, is not bound to a universe of discourse restricted to points in a space with a certain number of dimensions. The choice of a certain number of dimensions of a "space" determines certain relational predicates for the points, i.e., the individuals. Thus we have a certain model for predicate logic but not predicate logic of first order "for itself".

There is an assignment function which assigns individuals belonging to individual variables and, if admitted, constants to possible worlds. Regardless of how that is done the outcome will be bizarre if "world" is understood as a counterpart of the real world we live in and "individual" is understood as referring to the individuals we encounter in this world. However, if the world is understood as a set of state of affairs, nothing embarrassing happens. The individual variables for themselves are identities and they "bear properties", because there are assignment functions which assign them to sets. In classical logic, there is first of all the function which assigns the variable to the set selected by a property P. Classical logic is not a free logic. Thus there is also the tacit assumption that the variable is assigned to the set of identical entities which exist in the "actual world".

The foregoing considerations have shown that the sets of states of affairs, i.e., "worlds" in the sense of logic, do not have very much in common with the real world in which we live and are even different from the world "in itself" of science. Similarly it can be said that the individuals to which individual constants and individual variables refer do not have very much in common with the individuals of the real world, e.g., a stubnosed Socrates. It is "natural" for the "individual" represented by the individual variables to be assigned to or to be selected for sets of any sort. This belongs to its "nature". It belongs to the nature of Socrates only to the extent that it is possible to perform a formalising abstraction on something he has in com-

[22] Kant explicates this ontological theory of individuality and its roots in the framework of traditional logic in the *Critique of Pure Reason* (B 599ff). He himself does not share this view, which was taken up again by others, particularly Hegel.

mon with dungheaps and featherless chickens, namely that he is a "this here". There is no problem in assuming that the individuals which are "meant", "referred to" in a formalised language, can be assigned first to some set of state of affairs and then within that W; to the set which is selected by the predicate P. The very nature of a variable or a constant does not forbid that it be assigned to sets in other ways. Thus there is also no harm in assuming universe U to which all the individuals belong and letting them be connected with predicates in this universe and then be assigned to different sets of states of affairs.

The outcome of applying viewpoints taken from a phenomenological semantics on the level of predication to problems connected with the "possible worlds semantics" was that the "world" as it is to be understood in the context of formalised logic cannot be understood as some kind of concrete world, the real world in which we live, the "worlds" in quasi-positionality or worlds constructed in metaphysical speculations as correlates of other intellects. They are the result of a peculiar abstraction and idealisation which separates them in principle from all other conceptions of world. Thus the paradoxes which are connected with the attempts to connect the different concepts of "world" vanish. Quite another problem occurs which is a result of the separation. How can the explication of modal operators and other operators – as well as some intensional relations by means of this abstract apparatus – have objective validity for the explication of some categorical forms used in ordinary talk about the real world in which we live. Here we have a question about "objective validity" which belongs to transcendental logic. In the context of Husserl's phenomenology that means: it must be solved by means of an analysis of the genesis of these categorical forms in the prepredicative realm.

Thomas M. Seebohm was professor emeritus at the Johannes-Gutenberg Universität Mainz. For a more detailed description of his career and his work, please consult the "Introduction" to this volume.

Bibliography and Editorial Activity: Thomas A. Seebohm

Books

Monographs

1. Die Bedingungen der Möglichkeit der Transzendentalphilosophie, Bonn: Bouvier, 1961, 200 p.
2. *Zur Kritik der hermeneutischen Vernunft*, Bonn: Bouvier, 1972, 163 p.
3. *Ratio und Charisma. Ansätze zur Ausbildung eines philosophischen und wissenschaftlichen Weltverständnisses im Moskauer Rußland*, Bonn: Bouvier, 1977, 766 p.
4. *Philosophie der Logik*. Handbuch Philosophie, Freiburg: Alber, 1984, 364 p.
5. *Elementare formalisierte Logik*, Freiburg: Alber, 1991, 263 p.
6. Hermeneutics, Method and Methodology, Contributions to Phenomenology Vol. 50; Dordrecht/Boston/London, Kluwer Academic Publishers, 2004.

Editor

1. *Kant and Phenomenology*, Washington: Center for Advanced Research in Phenomenology and University Press of America, 1984. (Together with J.J. Kockelmans).
2. *Perspektiven Transzendentaler Reflektion, Festschrift für Gerhard Funke zum 75. Geburtstag*, Bonn: Bouvier, 1988. (Together with Gisela Müller).
3. Aron Gurwitsch, *Kants Theorie des Verstandes*, Contributions to Phenomenology Vol. 5, Dordrecht, Boston, London: Kluwer Academic Publishers, 1990.

© Springer Nature Switzerland AG 2020

T. Nenon (ed.), *Thomas Seebohm on the Foundations of the Sciences*,
Contributions to Phenomenology 105,
https://doi.org/10.1007/978-3-030-23661-8

4. *Phenomenology and the Formal Sciences*, Contributions to Phenomenology Vol. 8, Dordrecht: Kluwer Academic Publishers, 1991. (Together with Dagfinn Føllesdal and J.N. Mohanty).
5. *Prinzip und Applikation in der praktischen Philosophie*, AdWL Mainz, Stuttgart: Steiner, 1991.

Coeditor

1. *bewußt sein, Gerhard Funke zu eigen*, ed. A Bucher, H. Drüe, Th.M. Seebohm, Bonn: Bouvier, 1975,
2. *Continental Philosophy in America*, ed. H.J. Silverman, J. Sallis, Th.M. Seebohm, Pittsburgh: Duquesne University Press, 1983,
3. *Pragmatism Considers Phenomenology*, ed. R.S. Corrington, C. Hausman, and Th.M. Seebohm, Washington D.C.: Center of Advanced Research in Phenomenology and University Press of America, 1987,
4. *Proceedings of the 6. International Kant Congress, Vol. I, II/1, II/2*, ed. G. Funke and Th.M. Seebohm, Washington D.C.: Center of Advanced Research in Phenomenology and University Press of America, 1989.

Chapters in Books

1. "Lebenswelt", "Intersubjektivität", "Protention", "Quintilian", in: *Philosophisches Wörterbuch*, 16. Edition, Stuttgart: Kröner, 1961, pp. 279–80, 345–6, 479, 487.
2. "Wer ist ein Sohn des Vaterlandes?", A.N. Radiscev, In: *Aufklärung*, ed. G. Funke, Stuttgart: Koehler, 1963, pp. 363–9. (Translation).
3. "Erkenntnis aus dem Buch der Natur und aus der Heiligen Schrift", M.W. Lomonossow, In: *Aufklärung*, ed. G. Funke, Stuttgart: Koehler, 1963, pp. 316–22. (Translation).
4. "Reflexion, Interpretation und Dialektik", in: *Entwicklung und Fortschritt, Festschrift für W.E. Mühlmann*, ed. H. Reimann and E.W. Müller, Tübingen: J.B.C. Mohr, 1969, pp. 63–74.
5. "Bemerkungen zu Pazanin und Strasser", in: *Vérité et Vérification, Actes du Quatrieme Colloque International de Phénoménologie, Schwäbisch Hall, 8–11 septembre 1969*, ed. H.L. Van Breda, Den Haag: Nijhoff, 1974, pp. 183–4.
6. "Logik", in: *Lexikon der Pädagogik, Vol. III*, Freiburg: Herder, 1971, pp. 121–2.
7. "Das Widerspruchsprinzip in der Kantischen Logik und der Hegelschen Dialektik", in: *Akten des 4. Internationalen Kant-Kongresses Mainz, April 1974*, Ed. G. Funke, Berlin: de Gruyter, 1974, pp. 862–74.
8. "Zur Genese des Historismus", in: *bewußt sein, Gerhard Funke zu eigen*, ed. A Bucher, H. Drüe, Th.M. Seebohm, Bonn: Bouvier, 1975, pp. 111–24.

9. “On the Doctrine of Categories”, in: *Categories: A Colloquium*, ed. H.W. Johnstone, University Park, PA: Dept. of Philosophy, Penn State University Press, 1978, pp. 21–40.
10. “Historische Kausalerklärung”, in: *Kausalität: Neue Texte*, ed. G. Posch, Stuttgart: Philipp Reclam, 1981, pp. 260–88.
11. “The Significance of the Phenomenology of Written Discourse for Hermeneutics”, in: *Interpersonal Communication: Essays in Phenomenology and Hermeneutics*, ed. J.J. Pilotta, Washington D.C.: Center of Advanced Research in Phenomenology and University Press of America, 1982, 141–59.
12. “Die Kantische Beweistheorie und die Beweise der *Kritik der reinen Vernunft*”, in: *Akten des 5. internationalen Kant-Kongresses, Vol. II*, ed. G. Funke, Bonn: Bouvier, 1982, pp. 127–48.
13. “The New Hermeneutics, other Trends, and the Human Sciences from the Standpoint of Transcendental Phenomenology”, in: *Continental Philosophy in America*, ed. H.J. Silverman, J. Sallis, and Th.M. Seebohm, Pittsburgh: Duquesne University Press, 1983, pp. 64–89.
14. “The Other in the Field of Consciousness”, in: *Aron Gurwitsch in Memoriam*, ed. L.E. Embree, Washington D.C.: Center of Advanced Research in Phenomenology and University Press of America, 1984, pp. 283–304.
15. “Preface, Bibliography, Index”, in: *Kant and Phenomenology*, ed. Th.M. Seebohm and J.J. Kockelmans, Washington D.C.: Center of Advanced Research in Phenomenology and University Press of America, 1984, pp. V - XII, 203–229.
16. “Die Begründung der Hermeneutik Diltheys in Husserls transzendentaler Phänomenologie”, in: *Dilthey und die Philosophie der Gegenwart*, ed. E.W. Orth, Freiburg, München: Alber, 1985, pp. 97–1024.
17. “The End of Philosophy: Three Historical Aphorisms”, in: *Hermeneutics and Deconstruction*, ed. H.J. Silverman and D. Ihde, New York: New York University Press, 1985, pp. 10–23.
18. “Die Stellung der phänomenologischen Idee der Letzbegründung zur Seinsfrage”, in: *Einheit als Grundlage der Philosophie*, ed. K. Gloy and E. Rudolph, Darmstadt: Wissenschaftliche Buchgesellschaft, 1985, pp. 303–21.
19. “Phänomenologischen Betrachtungen zur Semantik möglicher Welten”, in: *Johannes Gutenberg Universität, Antrittsvorlesungen, Vol. 2*, Mainz: Universitätspressestelle, 1987, pp. 67–102.
20. “Wissenschaftsbegründung und Letztbegründung im Denkweg Martin Heideggers”, in: *Zur Selbstbegründung der Philosophie seit Kant*, ed. W. Marx, Frankfurt: Klostermann, 1987, pp. 157–75.
21. “Dilthey, Husserl, and Prima Philosophia”, in: *Dilthey and Phenomenology*, ed. R.A. Makkreel and J. Scanlon, Washington D.C.: Center of Advanced Research in Phenomenology and University Press of America, 1987, pp. 23–29.
22. “Considerations of a Husserlian”; In: *Pragmatism Considers Phenomenology*, ed. R.S. Corrington, C. Hausman, and Th.M. Seebohm, Washington D.C.:

Center of Advanced Research in Phenomenology and University Press of America, 1987, pp. 217–229.

23. "Foreword", in: G. Funke, *Phenomenology, Metaphysics, or Method*, Athens, Ohio, London: Ohio University Press, 1987, pp. VII - XV.
24. "Über die unmögliche Möglichkeit andere Kategorien zu denken als die unseren", in: *Kants transzendentale Deduktion und die Möglichkeit von Transzendentalphilosophie*, ed. Forum für Philosophie, Bad Homburg, Frankfurt: Suhrkamp, 1988, pp. 11–31.
25. "Über das Problem der Beschreibung einander bedingender Ereignisse", in: *Philosophie und Psychologie, Leib und Seele - Determination und Vorhersage*, ed. W. Marx, Frankfurt: Klostermann, 1989, 133–163.
26. "Bemerkungen zu einem Schriftenverzeichnis", in: *Perspektiven transzendentaler Reflexion, Festschrift für Gerhard Funke zum 75. Geburtstag*, ed. G. Müller and Th.M. Seebohm, Bonn: Bouvier, 1989, pp. 205–219.
27. "Kant's Theory of Revolution", in: *The Public Realm*, ed. R. Schürmann, Albany: State University of New York Press, 1989, 60–81.
28. "Apodiktizität, Recht und Grenze", in: *Husserl Symposion Mainz*, ed. G. Funke, Akademie der Wissenschaften und der Literatur Mainz, Stuttgart: Steiner, 1989, pp. 65–99.
29. "Transcendental Phenomenology", in: *Husserl's Phenomenology: A Textbook*, ed. J.N. Mohanty and W.R. McKenna, Washington D.C.: Center of Advanced Research in Phenomenology, Inc. and co-published by arrangement with the University Press of America, Inc., 1989, 345–385.
30. "The More Dangerous Disease: Transcendental Psychologism, Anthropologism and Historism", in: *Perspectives on Psychologism*, ed. M.A. Notturno, Leiden: E.J. Brill, 1989, pp. 11–31.
31. "Perspektiven des Lingualismus, Heidegger und Quine", in: *Martin Heidegger weiterdenken*, ed. A. Raffelt, Schriftenreihe der katholischen Akademie der Erzdiözese Freiburg, München, Zürich: Schell & Steiner, 1990, pp. 9–35.
32. "Vorwort des Herausgebers", in: Aron Gurwitsch, *Kants Theorie des Verstandes*, ed. Th.M. Seebohm, Contributions to Phenomenology 5, Dordrecht: Kluwer Academic Publishers, 1990, V - XX.
33. "Psychologism Revisited", in: *Phenomenology and the Formal Sciences*, ed. Th.M. Seebohm, D. Føllesdal and J.N. Mohanty Contributions to Phenomenology Vol. 8, Dordrecht: Kluwer Academic Publishers, 1991, pp. 149–182.
34. "The Paradox of Subjectivity and the Idea of Ultimate Grounding in Husserl and Heidegger", in: *Phenomenology and Indian Philosophy*, ed. D.P. Chattopadhyaya, L. Embree and J.N. Mohanty, New Delhi: Motilal Barnasidass Publishers, 1992, pp. 153–168.
35. "Possible Worlds", in: *Phenomenology - East and West (Festschrift Mohanty)*, ed. F.M. Kirkland and D.P. Chattopadhyaya, Dordrecht: Kluwer Academic Publishers, 1993.

36. "Über die vierfache Abwesenheit im Jetzt. Warum ist Husserl da, wo ihn Derrida nicht vermutet?", in: *Das Rätsel der Zeit, Philosophische Analysen*, ed. H.M. Baumgartner, Freiburg: Alber, 1993.
37. "Intentionalität und passive Synthesis", in: *Husserl in Halle*, ed. H.M. Gerlach and H.M. Sepp, Daedalus V, Frankfurt, Berlin et al.: Peter Lang, 1994.
38. "Considerations on *Der Satz vom Grund*", in: *The Question of Hermeneutics*, ed. T.J. Stapleton, Dordrecht: Kluwer Academic Publishers, 1994.
39. "Fichte's Discovery of the Dialectical Method", in: *Fichte, Historical Contexts/ Contemporary Controversies*, ed. D. Breazeale and T. Rockmore, New Jersey: Humanities Press, 1994.
40. "Kant und die Revolution", in: *Jahrbuch der Albertus-Universität zu Königsberg*, XXVIII, 1993, pp. 141–148.
41. "The Apodicticity of Absence", in: *Derrida and Phenomenology*, ed. W.R. McKenna and J.C. Evans, Dordrecht: Kluwer Academic Publishers, 1995.
42. "Some Difficulties in Kant's Conception of Formal Logic", in: *Proceedings of the 8. International Kant Congress, Memphis 1995*, Vol. I, Part 1, Section 1–2, ed. H. Robinson, Milwaukee: Marquette University Press, 1995.
43. "Literary Tradition, Intercultural Transfer and Cross-Cultural Conversations", in: *Cross-Cultural Conversations (Initiation)*, ed. A.N. Balslev, American Academy of Religion, Cultural Criticism Series Nr. 5, Atlanta, Georgia: Scholars Press, 1996, pp. 145–172.
44. "Kant und Mill über den Ursprung des obersten Prinzipes der Moral", in: *Inmitten der Zeit. Beiträge zur Europäische Gegenwartsphilosophie (Festschrift für Mafred Riedel)*, ed. Th. Grethlein and H. Leitner, Königshausen und Neumann, 1996, pp. 179–217.
45. "Individuals, Identity, and Names: Phenomenological Considerations", in: *Husserl in Contemporary Context*, ed. B.C. Hopkins, Dordrecht: Kluwer Academic Publishers, 1997, pp. 115–150.
46. "Johann Gottlieb Fichte", in: *Encyclopedia of Phenomenology*, ed. L. Embree et al., Dordrecht: Kluwer Academic Publishers, 1997, pp. 223–226.
47. "Germany" (together with E.W. Orth), in: *Encyclopedia of Phenomenology*, ed. L. Embree et al., Dordrecht: Kluwer Academic Publishers, 1997, pp.270–276.
48. "Hermeneutics", in: *Encyclopedia of Phenomenology*, ed. L. Embree et al., Dordrecht: Kluwer Academic Publishers, 1997, pp. 308–312.
49. "Logic", in: *Encyclopedia of Phenomenology*, ed. L. Embree et al., Dordrecht: Kluwer Academic Publishers, 1997, pp. 421–425.
50. "Vorwort" in: Gisela Müller, *Wahrnehmung, Urteil und Erkenntniswille. Untersuchungen zu Husserls Phänomenologie der vorprädikativen Erfahrung*, Bouvier, Bonn, 1999, pp. 5–19
51. "Die reine Logik, die systematische Konstruktion desd Prinzips der Vernunft und das System der Ideen" in: *Architektonik und System in der Philosophie Kants*, ed. H.F. Fulda/J. Stolzenberg, Meiner, Hamburg 2001, pp. 204–231.
52. "The Methodology of Hermeneutics as a Challenge for Phenomenological Research" in: The Reach of Reflection" in: *Issues for Phenomenology's Second*

Century, ed. S. Crowell, L. Embree, and S.J. Julian, www.electronpress.com, 2001, pp. 200–226.

53. "The Phenomenological Movement: A Tradition without Method? Merleau-Ponty and Husserl" in: *Merleau-Ponty's Reading of Husserl*, Ed. T. Toadvine and L. Embree, Kluwer Academic Publishers, Dordrecht/Boston/London 2002, pp. 51–68.
54. "The Hermeneutics of Texts. The Second Canon" in: *Hermeneutic Philosophy of Science, Van Gogh's Eyes, and God*, ed. B. E. Babich, Boston Studies in the Philosophy of Science 225, Kluwer Academic Publishers, Dordrecht/Boston/London 2002, pp. 137–152.
55. "Zum Problem des Verstehens" in: *Die Stellung des Menschen in der Kultur*, ed. C. Bermes, J. Jonas, K.-H Lembeck, Königshausen und Neumann, Würzburg 2002, pp. 123–143.

Journal Articles

1. "Zur Krise der Tradition unter Ivan III", in: *Zeitschrift für slavische Philologie* 32 (1965), pp. 108–123.
2. "Bemerkngen zur Klosterreform Josef Sanins", in: *Welt der Slaven* 14 (1969), pp. 430–450.
3. "Zwei neuere Explikate der Begriffe 'analytisch' und 'synthetisch'", in: *Kantstudien* 62 (1971), 202–217.
4. "Der systematische Ort von Herders 'Metakritik'", in: *Kantstudien* 63 (1972), pp. 59–73.
5. "Über die Möglichkeit konsequenzlogischer Kontrolle phenomenologischer Analysen", in: *Kantstudien* 63 (1972), pp. 237–246.
6. "Reflexion and Totality in the Phenomenology of Edmund Husserl", in: *Journal of the British Society of Phenomenology* 4 (1973), pp. 20–30.
7. "The Grammar of Hegel's Dialectic", in: *Hegel Studien* 11 (1976), pp. 149–80.
8. "Die Phänomenologie kognitiver Leistungen im Umgang mit formalen Sprachen", in: *Phänomenologische Forschungen* 2 (1976), pp. 49–75.
9. "Bemerkungen zum Problem der Interpretation irrealer Konditionalsätze als verkürzter Schlüsse", in: *Kantstudien* 68 (1977), pp. 1–17.
10. "The Problem of Hermeneutics in Recent Anglo-American Literature, Part I", in: *Philosophy and Rhethoric* 10 (1977), pp. 180–98.
11. "The Problem of Hermeneutics in Recent Anglo-American Literature, Part II", in: *Philosophy and Rhethoric* 10 (1977), pp. 263–75.
12. "Wertfreies Urteilen über fremde Kulturen im Rahmen einer transzendental-phänomenologischen Axiologie", in: *Phänomenologische Forschungen* 4 (1977), pp. 52–85.

13. "Reflection, Interpretation, and Dialectic", in: *Graduate Faculty Journal* 7 (1977), pp. 15–33.
14. "Schelling's Kantian Critique of Hegel's Deduction of Categories", in: *Clio* 8 (1979), pp. 239–55.
15. "Sowjetrussische Veröffentlichungen zum Kant-Jahr 1974", in: *Kantstudien* 70 (1979), pp. 491–507.
16. "Kant's Theory of Revolution", in: *Social Research: An International Quarterly of the Social Sciences* 48 (1981), pp. 557–587.
17. "Boeckh and Dilthey: The Development of Methodical Hermeneutics", in: *Man and World* 17 (1984), pp. 325–46.
18. "Fichte's and Husserl's Critique of Kant's Transcendental Deduction", in: *Husserl Studies* 2 (1985), 53–74.
19. "Facts, Words, and what Jurisprudence Can Teach Hermeneutics", in: *Research in Phenomenology* 16 (1986), pp. 24–40.
20. "Isiodor of Seville versus Aristotle in the Questions on Human Law and Right in the *Summa Theologiae* of Thomas of Aquinas", in: *Graduate Faculty Journal* 11 (1986), pp. 83–106.
21. "Deconstruction in the Framework of Methodical Hermeneutics", in: *Journal of the British Society of Phenomenology* 17 (1986), pp. 275–88.
22. "Phenomenology of Logic and the Problem of Modalizing", in: *Journal of the British Society of Phenomenology* 19 (1988), pp. 235–251.
23. "Kategoriale Anschauung", in: *Phänomenologische Forschungen* 23 (1990), pp. 9–43.
24. "Falsehood as the Prime Mover of Hermeneutics", in: *The Journal of Speculative Philosophy NS* 6 (1992), pp. 1–24.
25. "Variable, Objekte, Mengen von Universen und maximale Konsistenz in formalisierten Sprachen" (Metakritik zur Diskussion von L.B. Puntel), in: *Ethik und Sozialwissenschaften, Streitforum für Erwägungskultur* 3 (1992), pp. 186–195.
26. "The Pre-conscious, the Unconscious, and the Subconscious: A Phenomenological Explication", in: *Man and World* 25 (1992), pp. 505–520.
27. "The Pre-conscious, the Unconscious, and the Subconscious: A Phenomenological Critique of the Hermeneutics of the Latent", in: *Aquinas. Rivista Internazionale di Filosofia* 35 (1992), pp. 47–271. (Extended version of 26)
28. "L'Individuo. Considerazioni fenemenologiche su una categiria logica", in: *Discipline Filosofiche* (1993.1), pp. 21–71.
29. "Logika ponjatii kak predosylka kantovsoj formal'noj i transcedental'noj logiki" (translated by V. Brijuschinkin), in: *Kantovskij Sbornik* 17 (1993).
30. (T.M. Zeboms), "Parvertejot psihologismu" transl. from the English (Bibl. Nr.) by E. Picukane, in: *Kentaurs XXI*, Riga, 2001, pp. 23–50.

Reviews

1. "*Beiträge zu Philosophie und Wissenschaft*, ed. E Höfling", in: *Archiv für Rechts- und Sozialphilosophie* 48 (1962), pp. 265–9.
2. "H. Drüe, *Edmund Husserls System der phänomenologischen Psychologie*", in: *Zeitschrift für philosophische Forschung* 18 (1963), pp. 523–31.
3. "K. Löwith, *Die Hegelsche Linke*", in: 49 (1963), pp. 119–25.
4. "St. Buczkowski, "Prawo a problemy ekonomiczne", W. Wolter, "Z roswazan-nad wina nie umyslna", A. Podgorecki. "Sociologia a nauki prawne", in: *Panstwo i Prawo* 17 (1962)", in: 49 (1963), pp. 360–5.
5. "*Hegel bei den Sklaven*, ed. D. Tschyzevskij", in: *Zeitschrift für Ostforschung* 17 (1968), pp. 127–32.
6. "R. Sokolowski, *Husserlian Meditations*", in: *Philosophische Rundschau* 23 (1976), pp. 60–5.
7. "G. Planty-Bonjour, *Hegel et la pensée philosophique en Russie 1830 - 1917*", in: *Hegel Studien* 11 (1976), pp. 313–6.
8. "A. Walicki, *The Slavophile Controversy*", in: *Hegel Studien* 12 (1977), pp. 258–61.
9. "H.G. Gadamer, *Philosophical Hermeneutics*", in: *Philosophy and Rhethoric* 11 (1978), pp. 191–5.
10. "P. Ricoeur, *Interpretation Theory: Discourse and the Surplus of Meaning*", in: *Graduate Faculty Philosophy Journal* 7 (1978), pp. 257–69.
11. "A. Gulyga, *Kant*", in:*Kantstudien* 70 (1979), pp. 234–6.
12. "Hegel, *Gesammelte Werke*, Vol. 11, ed. F. Hogemann and W. Jaeschke", in: *Clio* 11 (1981), pp. 440–4.
13. „Elfriede Conrad, *Kants Logikvorlesungen als neuer Schlüssel zur Architektonik der Kritik der reinen Vernunft.* Forschungen und Materialien zur Deutschen Aufklärung. Stuttgart/Bad Cannstatt: Frommann-Holzboog, 1994 "in *Journal of the History of Philosophy* 1996, pp. 620–21

Editorial Boards

Series

1. *Current Continental Research*, University Press of America
2. *Series in Continental Thought*, Ohio University Press
3. *Contributions to Phenomenology*, Kluwer Academic Publishers

Journals

1. *Kantstudien*
2. *Husserl Studies*
3. *Graduate Faculty Studies*
4. *Man and World*
5. *Journal of Speculative Philosophy, NS*
6. *Allgemeine Zeitschrift für Philosophie*
7. *Phänomenologische Forschungen*
8. *Journal of the History of Philosophy*

MIX
Papier aus verantwortungsvollen Quellen
Paper from responsible sources
FSC® C105338

If you have any concerns about our products,
you can contact us on
ProductSafety@springernature.com

In case Publisher is established outside the EU,
the EU authorized representative is:
Springer Nature Customer Service Center GmbH
Europaplatz 3, 69115 Heidelberg, Germany

Printed by Libri Plureos GmbH
in Hamburg, Germany